# ON-DEMAND PRINTING
## *The Revolution in Digital and Customized Printing*

Second Edition

**Howard M. Fenton**
*Graphic Arts Technical Foundation*

**Frank J. Romano**
*Rochester Institute of Technology*

**Graphic Arts Technical Foundation**
Pittsburgh, Pennsylvania

Graphic Arts Technical Foundation
200 Deer Run Road
Sewickley, PA 15143-2328
Phone: 412/741-6860
Fax: 412/741-2311
http://www.gatf.lm.com

# Table of Contents

# Dedication

I would like to dedicate this book to three mentors.

My father asked me to work after school, weekends, and summers in his business. Working with dad, I learned about business and the importance of the entrepreneurial spirit.

Jeff Liebman hired me as a lab assistant and promoted me to a scientist. Working with Jeff, I learned about the scientific method and neuroscience research. After a few years I had several first-authored publications and presentations.

Frank Romano hired me to write for *TypeWorld* and make presentations at Type-X. After working with Frank, I became the editor of *Pre* and was invited to present over 100 seminars a year.

For all my mentors — thanks.

Howard Fenton

# Acknowledgments

Thanks to all the manufacturers who supplied artwork for the book.

Special thanks to Kodak and Agfa for providing most of the graphics.

Thanks to Frank V. Kanonik, director of on-demand printing services at GATF, for his contributions to the DocuTech and bindery chapters.

Thanks to Jack Powers, Mary Lee Schneider, and George Alexander who reviewed an early version of this manuscript and had suggestions that resulted in a better book.

Thanks to Melinda Bateman, Barb Pellow, and Cary Sherburne for help with the second edition.

Thanks to Ron Krivosheiw, president of Speed Graphics, for the opportunity to participate in the management and the opportunity to work "hands-on" in one of the largest on-demand printing facilities, during a consulting assignment.

Thanks to Tom Destree, editor in chief, who edited the manuscript, created many of the line illustrations, and paginated the entire book, and to Charles Lucas, who created the cover design.

# Introduction

The printing industry is changing. Technology is being applied to increase the productivity of traditional processes, while at the same time other technology is challenging those processes. The concepts of on-demand printing can be applied to conventional printing, but, more likely, they will be applied to digital printing.

This publication summarizes the concepts, technologies, and applications of on-demand printing. It will help you to understand the rapid changes that are taking place and provide a base of knowledge for decision making.

GATF has been involved with on-demand approaches since its installation of a Xerox DocuTech in 1991. Many of our publications are now produced on-demand. Using this approach, we have minimized our inventory costs and increased the frequency of revisions to our textbooks.

The market for short-run print production is expanding, and customers now define turnaround time in hours rather than days. The technology continues to evolve, incorporating color into the many on-demand solutions on the market.

GATF is committed to assist its members in sorting through the solutions. We will continue to evaluate the technology using our test forms and producing our short-run textbooks. We will communicate our experiences through our publishing efforts.

We believe that on-demand printing presents many opportunities to the printing and publishing industry to expand their capabilities for their customers. Our experience has indicated that it provides a nice complement to the lithographic process and is an alternative to electronic distribution of data.

George Ryan
President
Graphic Arts Technical Foundation

# Section I

# Markets and Applications

# Chapter 1

# What Is On-Demand Printing?

Whenever we speak to an audience about on-demand, digital, or custom printing, an argument always ensues over what exactly we are talking about. Apparently, there is no consensus of opinion for the definitions of on-demand printing, on-demand presses, digital printing, and digital presses. They sound the same, but they are not.

For example, the category of on-demand presses contains the DocuTech and Lionheart, which are electrophotographic or copier-based technologies. Another controversial issue is how to categorize the Heidelberg GTO-DI? Is it a digital press? Is it an on-demand press?

And possibly the most controversial: is it ever possible for a conventional press to perform on-demand printing, digital or custom printing or publishing? Before we go any further, let's stop and define these terms.

## Defining "On-Demand"

In a generic sense, the concept of on-demand is basically one of short notice and quick turnaround. Therefore, if you want a movie or fax on-demand, you simply push a button or make a phone call — you give short notice and get quick turnaround.

In the printing industry, it is also associated with shorter and usually more economical printing runs. When all of this is combined, the definition becomes "short notice, quick turnaround of short, economical print runs." When all criteria are met, it results in lower inventory costs, lower risk of obsolescence, lower production costs, and reduced distribution costs.

In general, most traditional printing does not fulfill this criteria and does not result in these advantages. That is why the disadvantage of traditional long-run printing is that the reproduced information can become out-of-date, which requires the disposal and re-manufacture of new material. In the United States, it is estimated that 31% of all traditional printing is discarded because it becomes obsolete. This number is composed of 11% of all publications, 41% of all promotional literature, and 35% of all other material.

Magazine publishing is a good example of this. Approximately one-third of all magazines displayed on a newsstand today are discarded. Many books are never sold, and some are even discarded by the bookseller. Another good example are the stories told about the disposal of obsolete forms by forms manufacturers: tractor trailers filled with forms or data/specifications sheets that go to a landfill or dump because of product changes. The relentless push of technology causes products to be changed more frequently.

Notice that although we say that most traditional printing does not fulfill this criteria, we do not associate any particular technology with the concept of on-demand. If you are clever or efficient or high tech enough, you could produce on-demand printing with a traditional press.

According to our definition it doesn't matter what technology you use. The customer probably doesn't care. It's rare for a customer to ask if you printed such-and-such on a Heidelberg or a MAN Roland or did you copy this on a Canon or a Xerox machine. They don't really care. As long as the quality is acceptable, and it is done quickly and economically, they will be happy.

In addition, most customers today use the terms "printing" and "copying" interchangeably, without any thought as to what they signify. When someone talks about an on-demand press, they generally are referring to a Xerox DocuTech or Kodak Lionheart, which use electrophotographic or copying technology.

For our intents and purposes, an on-demand press is any device that can print short runs, on short notice, relatively quickly, in a cost-efficient manner. This can be accomplished with a traditional press, a high-speed copier, a hybrid technology press, a high-quality printer, or a color copier. Lastly, the terms "on-demand printing" and "demand printing" will be used interchangeably.

## Defining Digital Printing

The definition of digital printing is a bit more difficult. A simple definition, and the one that we will use throughout the book, is that digital printing is any printing completed via digital files. Two points to note — first, we avoid discussing specific technologies, and second, when referring to digital printing, the short-run aspect is eliminated.

A digital press may be capable of printing short runs economically, but digital printing on printing presses is well-suited for slightly longer runs. For example, a Heidelberg GTO-DI is definitely a digital press, but it may or may not be an on-demand press.

Demand printing is economical, fast, and oriented to short runs while digital printing is printing from digital files, but is not restricted to short runs. Demand printing can be done with digital files or conventional film or plates, while digital printing is done only with digital files.

The advantage of this definition is that it deals with the Heidelberg GTO-DI — a controversial press. Some say that a traditional press is one that uses a static image, while a digital press is one that uses a variable image. With this definition, the GTO-DI would not be a digital press. Some say it should be considered a demand press because it can print out 5,000 impressions quickly and economically, but others say that it should not because it cannot perform variable printing.

Our goal in this book is to cover all aspects of demand printing, with an emphasis on the digital presses that use new technology. But, you can see that we are trying to cover a larger area — namely, demand printing with any reproduction system. The next terms to define are "variable information," "variable printing," and their evil cousin — "static printing."

## Defining Variable Printing

Traditional printing does not allow you to print variable information. With traditional printing, the prepress work is performed, the plates are made, and they are run on the press. The end result is thousands of pages that look exactly the same. This information is static and, therefore, not variable.

In contrast, many of the digital presses (presses that print from digital data) can print variable information. On different pages you could have different names and addresses. The ability to print variable information which results in variable printing is the critical component of customized printing.

To accomplish customized printing today, conventional pages or static pages are run through high-speed inkjet devices for the variable information. This is how Publishers Clearinghouse Sweepstakes sends you a package that says "Congratulations 'your name' — you have been picked as one of our finalists." This is a simple example of how variable information can be used to create a personalized or customized message.

Although traditionally accomplished with high-speed inkjet devices, many of the digital presses offer this ability. For example the Xerox DocuTech, Indigo E-Print 1000, Xeikon DCP-1, and Agfa Chromapress can take information from a database and create different information on different pages. An added advantage is that these devices are not limited to six or twelve lines of copy, as many of the inkjet devices are, but some can customize the entire page. For many, this is one of the most exciting elements of these devices.

These abilities allow you to vary the contents of any single document or group of documents during a run. For instance, a cataloger could print 5,000 catalogs in lots of 500, each with a different cover. This could be used to test-market different covers. This would have been almost impossible or very cost prohibitive with traditional techniques.

There are as many ways to use variable information as your imagination will allow. The cataloger could include the recipient's name on each of the 1,000 direct mailers and vary a special insert based on demographics. The cataloger could also personalize sweatshirts and offer "iron-on" names. R.R. Donnelley and Sons, Inc. has had tremendous success with this type of personalization.

The building blocks of customized printing is the combination of variable information with output devices that do not require intermediate films or plates. They are true digital printing systems in that all or part of the image area can be changed from impression to impression.

In describing printing devices capable of variable printing, we must discuss technologies. As described above, only digital presses allow the use of variable information. A press with a traditional printing plate cannot print customized data or incorporate variable data without a plate change, although customized printing can be accomplished with high-speed inkjet printing.

If you look at the amount of mailings you receive where your name is built into the message, you will get some idea of the tremendous market for personalization and customization of text and images. The objective of any direct mail piece is to encourage you to read it. Personalization accomplishes this because you are more apt to read something that grabs your interest — and what grabs your interest more than YOU?

## Our Definition

Our definition of a printing process capable of variable printing is one that can incorporate data from a database and does not use analog technologies such as film or plates.

How does the definition of variable printing interact with previous definitions? Completely independently. A demand press may or may not be able to print variable pages, and a digital press may or may not be able to print variable pages.

With these definitions, a traditional press that performs demand printing but not variable printing and a digital press that does not print variable information (GTO-DI) can exist. Confused? The following chart may clear things up.

| Device Category | Run Length | Advantages | Variability |
|---|---|---|---|
| Demand press | Short run | Quick turnaround, economical | No or yes |
| Digital press | Any run | Accept digital file | No or yes |
| Variable on-demand | Short run | Quick turnaround, economical | Yes |
| Variable digital press | Any | Accept digital file | Yes |

Of course, there are presses that are demand and digital presses that print out variable information. Traditional presses can therefore perform demand printing, and digital presses or digital printing may not use variable information.

## How Short Is Short Run

Actually, there are different degrees of run length ranges, according to GAMA:

| | | | |
|---|---|---|---|
| 1 | Ultimate Short Run (USR) | 6501–8000 | Moderate Run (MR) |
| | | 8001–10000 | |
| 2–100 | Ultra Short Run (VVSR) | 10001–25000 | |
| | | | |
| 101–250 | Very Short Run (VSR) | 25001–40000 | Average Run (AR) |
| 251–500 | | 40001–50000 | |
| 501–1000 | | 50001–70000 | |
| | | | |
| 1001–1500 | Short Run (SR) | 70001–100000 | Moderate-Long Run (MLR) |
| 1501–2000 | | 100001–250000 | |
| 2001–3000 | | 250001–400000 | |
| | | | |
| 3001–4000 | Moderate-Short Run (MSR) | 400001–600000 | Long Run (LR) |
| 4001–5000 | | 600001–750000 | |
| 5001–6500 | | 750001–1000000 | |
| | | | |
| *Source: GAMA* | | >1,000,000 | Very-Long Run (VLR) |

## What Are Typical Lengths

The best definition of a short run is one that is less than 5,000 impressions. The next question is, how much work is done at this run length? Many are surprised at the answers. Almost 56% of commercial, book, and office printing including duplicating and copying falls in the category of run lengths from 500 to 5,000 impressions.

It is interesting that only 2.8% of this printing is done in four or more colors. But, these numbers represent the trends today. According to the market research discussed later, by the year 2000 the amount of four-color printing in this run-length market will more than quadruple to 11.5%. In fact, color will increase as a percentage of total reproduced pages as it becomes easier to accomplish on new and traditional equipment.

Traditional printing presses generally operate in the long-run category and above. Recently, however, many printers have been attempting to compete with moderate or even short-run categories to remain competitive.

Technological advances that can be applied to traditional presses, such as quick-change plate capability, on-press densitometry, calibrated adjustments, and waterless offset, could move some presses into the short-run category routinely. That is why we have left the definition of demand printing open to any technology. It is possible for clever printers to accomplish on-demand printing with traditional presses.

Most traditional presses as we know them now, however, will have difficulty in meeting the needs of the short-run category and below. This is where a new market is developing. Based on market projections and our conversations with printers, we project that the short-run wars will take place in the 100- to 3,000-copy range. Above that, newer printing presses will be competitive, and below that, newer plain-paper color printers will evolve.

| Run Lengths | % of Total Market | % of Run Length in 4-Color | % Total 4-Color Market | |
|---|---|---|---|---|
| | | | 1994 est. | 2000 est. |
| <100–500 | 16.6 | 1.0 | 0.2 | 3.5 |
| 500–2,000 | 33.5 | 3.0 | 1.0 | 5.5 |
| 2,000–5,000 | 22.3 | 10.0 | 1.8 | 6.0 |
| 5,000–10,000 | 13.8 | 16.0 | 2.2 | 5.0 |
| 10,000–100,000 | 5.6 | 25.0 | 1.4 | 1.5 |
| >100,000 | 8.2 | 41.0 | 3.4 | 5.5 |
| **Total %** | **100.0** | | **10.0** | **27.0** |

## Short-Run Process Color Printing

In defining demand printing, we stated that short runs, on short notice, relatively quickly, in a cost-efficient manner. But what exactly is a short run? Different companies that use different printing technologies define short runs differently. What might be a short run for gravure, might be considered a long run for sheetfed.

At the Agfa Technology Expo, Larry Chism, business manager of the Moore Research Center, discussed the topic of run length. Second only to R.R. Donnelley and Sons, Chism has been beta testing digital presses longer than anyone in the United States. He has worked with the Xerox DocuTech, the Indigo E-Print 1000, and Xeikon DCP-1 digital printing presses.

According to Chism, run lengths from 100 to 1,500 are ideal for digital color presses. Beyond that, he said, it is difficult to turn a profit. Digital printing allows Moore to serve its customers by storing multiple-page documents that are needed in small quantities, on-demand.

It also allows Moore to offer full variable color printing in which every document can be different or have a personal message. For more information about how commercial printers are implementing on-demand services, see the chapter dedicated to the subject.

## On-Demand Printing and Publishing Concepts

In an oversimplification of the on-demand process, the client supplies electronic files or camera-ready materials and specifies how many copies of the publication will be needed. The printer produces the publication directly from the disk or camera-ready artwork and delivers it within a specified timeframe.

There are many aspects to the process that make it more than simple. With short-run work, it is necessary to automate the job submission so that all information about the job accompanies the actual electronic file. This is only one aspect of on-demand that escapes potential users. Short runs imply that more jobs will be passing through a facility, with average runs of under 1,000. This will force the modification of workflow and billing procedures.

Currently there are three specific on-demand strategies in our industry—on-demand printing, distributed demand printing, and on-demand publishing

Except for rare occasions when traditionally prepared and printed short runs are fast and economical, most of the time the expression "on-demand" means that the data is stored and printed in electronic form. It doesn't have to be an electronic file, but generally it is a digital file which facilitates the efficacy of the short run. (Remember, any reproduction process can be on-demand or short run.)

The second strategy is known as distributed demand printing. Unlike the general concept of demand printing, the distributed demand printing workflow requires that the electronic files can be transmitted to other locations, printed, and distributed locally. These publications can then be stored, printed, and shipped locally as needed.

This an implementation of the distribute-and-print philosophy as opposed to the traditional print-and-distribute philosophy. The traditional long-run printing strategy is to print large volumes in a central location and then ship them both long (nationally) and short distances (regionally). Decentralization does reduce shipping costs but does not eliminate storage and distribution costs. Combining on-demand printing with decentralization produces the best results.

The third general strategy is demand publishing in which the data is stored in paginated form and transmitted for immediate printout. Large-volume magazines do this—which allows them to provide regional inserts. Portable Document Formats, such as Adobe Acrobat or No Hands Common Ground, are being used to distribute the paginated and print-ready page and document files.

Another example of this is fax publishing, which makes every fax machine a printing press. Paginated and formatted files are transmitted for printout. Quality is still an issue, but over time any printer may be fax-like and resolutions of 600 dpi or more would be possible.

## Advanced Definition of Digital Printing

Although not the subject of this book, a more advanced definition of digital printing would be *any printing that uses a rasterization process to produce image carriers or to replicate directly to substrate from digital document files.* The chart on the next page details many alternatives for getting page images to paper, either via an image carrier or directly to a substrate.

Thus, we include direct-to-plate (and stencil) technology in our definition, as well as direct-to- imposed film (not shown in the chart). Imposed film is one of the steps to produce a plate, of course.

Eventually, digital approaches will apply to just about every facet of graphic communication, whether it applies to a run of 1,000, 100, or just one.

## Future On-Demand

In the future, our definition of on-demand printing may change to include binding or finishing. Ultimately, on-demand printing requires both an imaging engine and a means of combining in consecutive, uninterrupted operations the printed pages into finished products — college textbooks, out-of-print books, insurance policies, research reports, business proposals, or any other reproduced products.

Printing with in-line finishing puts very stringent demands on the condition and reliability of the equipment used. When any part of the line is down, the whole system is down. Also, printing and finishing require different skills and operators will require both kinds.

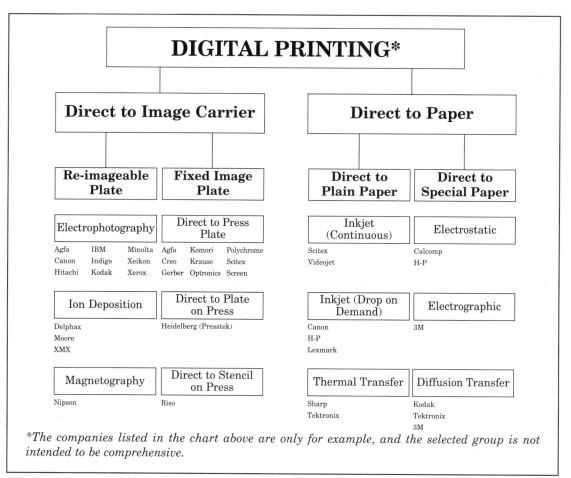

**DIGITAL PRINTING\***

| **Direct to Image Carrier** | | **Direct to Paper** | |
|---|---|---|---|
| **Re-imageable Plate** | **Fixed Image Plate** | **Direct to Plain Paper** | **Direct to Special Paper** |
| Electrophotography | Direct to Press Plate | Inkjet (Continuous) | Electrostatic |
| Agfa  IBM  Minolta<br>Canon  Indigo  Xeikon<br>Hitachi  Kodak  Xerox | Agfa  Komori  Polychrome<br>Creo  Krause  Scitex<br>Gerber  Optronics  Screen | Scitex<br>Videojet | Calcomp<br>H-P |
| Ion Deposition | Direct to Plate on Press | Inkjet (Drop on Demand) | Electrographic |
| Delphax<br>Moore<br>XMX | Heidelberg (Presstek) | Canon<br>H-P<br>Lexmark | 3M |
| Magnetography | Direct to Stencil on Press | Thermal Transfer | Diffusion Transfer |
| Nipson | Riso | Sharp<br>Tektronix | Kodak<br>Tektronix<br>3M |

*\*The companies listed in the chart above are only for example, and the selected group is not intended to be comprehensive.*

# Chapter 2

# Market Research

One of the forces driving the interest and excitement in on-demand, digital, and variable printing are market research reports. There are several companies that specialize in market projections in the graphic arts, such as BIS Strategic Decisions in Norwood, Mass.; Charles A. Pesko Ventures (CAPV) in Marshfield, Mass.; and State Street Consulting in Boston, Mass. Although each company and study is different, they all come to the same conclusion — the market is growing, and growing fast.

According to State Street Consulting, the short-run printing market is $16 billion and will grow to $28 billion by the year 2000. This could, however, be an underestimation of the interest this technology will generate. It is only a market estimate of the amount of printing sales; it is not an indication of the number of sites that may consider this technology, and probably does not put a value on in-house work.

Charles Pesko believes that digital color presses will transform printing from a craft into a service. He refers specifically to on-demand printing, which he defines as: *the ability to get what you want when you want, and where you want it.* In 1993, on-demand services accounted for $7.2 billion, or 9%, of the $80.2 billion commercial printing market. Pesko predicts that by the year 2000, on-demand will represent 20%, or $16 billion, of that market.

It is not surprising that nearly half of the commercial printing market consists of print runs under 5,000, of which 20% ($7.5 billion) is produced in full color. These numbers will increase rapidly as digital color printing devices reach the market. Pesko believes that five things — "convenience, easy updates, personalization, reduced inventory, and in-line bindery" — will open the door to new digital opportunities.

According to the Graphic Arts Marketing Information Service (GAMIS) report entitled *The Impact of Quick Printing on the Commercial Printing Industry,* the fast-turnaround, short-run market will grow faster then any other market segment. The study claims that this market is currently 25% of the total market and will grow to 30% by the year 1997. The authors CAPV (Charles A. Pesko Ventures) also projected that nearly half of all printing done in the future will be in short runs. In other published research, CAPV projects that 47% of the $80 billion market will be short run. This market will be composed of 43% one-color, 37% two- and three-color, and 20% four-color printing.

## A Market Subset

Short-run printing is a subset of the printing market in general. It is not that buyers want only short-run printing; they want printing, and some of it will be short-run printing.

We must do more to define what we do as an industry, not how we do it. We use the terms printing and copying without any thought as to what they signify. It is said that we define ourselves by our biggest machine, and that is a printing press. Copying involves machines that take an image and reproduces it without traditional plates or inks. Printing implies ink; yet, digital printers do not use traditional ink.

Our point is that there has always been a short-run, on-demand market. We applied copiers and electronic printers to that market, and will continue to do so. Now, however, we are applying specially equipped offset lithographic presses and a new breed of digital presses.

Xerox describes the on-demand market:
- 3.3 billion pages in North America (N.A.) and Europe in 1993
- $114 billion in market value

A leading consultant says:
- 1000 and under run length segment = $14 billion (N.A.) at cost
- Single color: 55%
- Two and three color: 30%
- Process color: 15%

Another leading consultant says:
- 113 billion pages in North America alone
- $26 billion in market value

Here are the CAPV projections in chart form; however, they differ greatly from the previous projections:

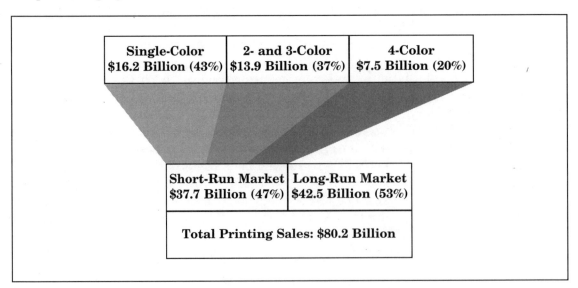

| Single-Color $16.2 Billion (43%) | 2- and 3-Color $13.9 Billion (37%) | 4-Color $7.5 Billion (20%) |

| Short-Run Market $37.7 Billion (47%) | Long-Run Market $42.5 Billion (53%) |

Total Printing Sales: $80.2 Billion

At the Digital Color Printing and Electronic Printshop seminar presented by BIS Strategic Decisions, it was stated that 55% of all pages — both offset and copied — were printed in one color, 38% in two colors, and 6.8% in four color during 1993. Of the pages produced electrophotographically (copied) 84% were one color, 15% two color, and 0.5% full color. They project that by 1998 full-color pages are expected to claim 22% of the short-run market, two-color pages 30%, and single-color pages 48%.

An article in *Color Publishing* magazine (Jan/Feb 1995) reported that the BIS Strategic Decisions Group projected that the price of digital presses will fall dramatically until the end of 1995, while revenue produced by these products will not exceed the cost until 1996. They project that the average entry cost for digital presses will be $300,000, the sales based on the products will be 600 in 1996, and sales will grow to $1,100,000 by 1998.

## Whose Work Will Change?

With these tremendous shifts in print production from longer runs to shorter runs, some important questions remain:
- Who is doing the work today?
- Who will do the work tomorrow?
- Who should be concerned about work changing?
- Which work is best suited for on-demand printing and publishing?

In an attempt to answer some of these questions, a number of tables are included. These tables contain information about the types of work done, who is doing it, and which may be best suited for on-demand solutions based on run lengths. The table below shows the various categories of print production and where that work is done today.

| Work Type | Commercial Printing | Commercial Copier/Dupl. | In-house Printing | In-house Copier/Dupl. | Office Printer |
|---|---|---|---|---|---|
| Ad/Direct Mail | 79% | 9% | 6% | 2% | 4% |
| Flyer/Folder | 91% | 3% | 2% | 3% | 1% |
| Brochure/Booklet | 67% | 16% | 1% | 10% | 6% |
| Newsletter | 58% | 14% | 8% | 19% | 1% |
| Book, Directory | 99% | 1% | — | — | — |
| Magazine, Journal | 95% | 1% | 3% | 1% | — |
| Catalog | 95% | 1% | 2% | 1% | 1% |
| Financial, Legal | 89% | 4% | 1% | 5% | 1% |
| Form/Coupon | 54% | 7% | 22% | 16% | 1% |
| Sign/Poster | 96% | 3% | 1% | — | — |
| All Other | 39% | 22% | 16% | 14% | 9% |
| **Average** | **78%** | **7%** | **6%** | **7%** | **2%** |

*Source: GAMA interviews with printers and print buyers*

## Where Are Pages Created?

The following chart separates page design into those created in an office environment, and those created commercially or by trade professionals such as typesetters, service bureaus, or printers. They are also separated into those printed in-house, versus those printed commercially.

| Created<br>Reproduced | *Office*<br>*In-plant* | *Office*<br>*Commercially* | *Commercially*<br>*In-plant* | *Commercially*<br>*Commercially* |
|---|---|---|---|---|
| DOCUMENT | | | | |
| Internal | 20% | 69% | — | 11% |
| External | 6% | 62% | 5% | 27% |
| | | | | |
| PUBLICATION | | | | |
| Internal | 25% | 52% | 11% | 12% |
| External | 11% | 38% | 9% | 42% |
| | | | | |
| PROMOTIONAL | | | | |
| Advertising | 1% | 14% | 6% | 79% |
| Direct mail | 1% | 30% | 1% | 68% |
| | | | | |
| UTILITY | | | | |
| | 10% | 49% | — | 41% |
| | | | | |
| Average | 11% | 45% | 4% | 40% |

*Source: Interviews with commercial printers, in-plant printers, and print buyers*

According to the numbers in the chart, 56% of all pages are not created by trade professionals, but 85% are reproduced (printed) there. This shows the dramatic changes that have occurred with desktop technology. Until about 1987, most, if not all pages were produced commercially because specialized prepress technology was required and not readily available except in commercial print shops. Printers today report that they receive an increasing percentage of their work electronically. In 1995, the number of pages received in electronic form was approximately 57%, while in 1999 that number is projected to increase to 81%.

## Number of Originals and Run Length

One important and complicated question is what type of work is best-suited for on-demand printing. One determining factor is the number of original pages (or pages per unit) and the number of reproductions required. A flyer may be only one sheet folded to a certain format, a book is usually perfect bound, and magazines are often saddle-stitched. The document must be considered a finished unit in order to apply the principles of on-demand printing.

## Reproduction Run Length Versus Page Count Per Document

The following table works in several ways. It describes the relationship between the number of originals per job, and the number of copies of that unit that are reproduced. For instance, the first column at left shows that 21% of all documents and publications in the U.S. are under 10 pages. As you move right, the table shows that 13% of those under-10-page units are reproduced with under 100 copies.

|  | Pages per unit | Copies per unit | | | | | | | |
|---|---|---|---|---|---|---|---|---|---|
|  |  | <100 | 101–500 | 501–1500 | 1501–2000 | 2001–5000 | 5001–10M | 10M–100M | >100M |
| 21.1% | Under 10 pages | 13% | 14% | 11% | 6% | 9% | 31% | 6% | 10% |
| 19.9% | 11–20 pages | 11% | 9% | 26% | 17% | 19% | 8% | 3% | 7% |
| 14.1% | 21–50 pages | 2% | 11% | 21% | 11% | 29% | 7% | 10% | 9% |
| 14.5% | 51–100 pages | 7% | 12% | 16% | 19% | 17% | 13% | 10% | 6% |
| 14.1% | 101–200 pages | 3% | 9% | 17% | 21% | 29% | 10% | 1% | 10% |
| 16.3% | Over 201 pages | 2% | 6% | 14% | 22% | 31% | 14% | 4% | 7% |
| 100.0% |  | 6.4% | 10.2% | 17.5% | 16.0% | 22.4% | 13.8% | 5.6% | 8.2% |

*Source: GAMA interviews, demand printing report*

There appears to be no pattern to the number of pages per unit in terms of radical departures from a simple average. Our categories of page count average 16.6% each with a low of 14.1% and a high of 21.1%. In other words, neither small nor large page counts dominate. Likewise, copies per unit show no major deviation. However, Very Short Run (VSR) and Short Run (SR) jobs total 50.1% and the addition of Moderate Run (MR) jobs brings the total to 86.3%.

## New Technologies Shift Existing Methods

Studying past effects of new technology on previous communication methods, one thing becomes perfectly clear: new technologies change consumer demand and consumption. Let's use two examples — Automated Teller Machines (ATMs) and television sets (TVs).

It wasn't that long ago that you had to go to a bank (and only your bank) or the corner grocery store to cash a check and get money. The emergence of a new technology, the ATM, changed that. Today, it only takes a plastic credit card and a personal identification number (PIN) to get money at any bank in the United States or abroad.

This is a good example of how new technologies change consumer demand. Today, you probably would not even consider opening a checking account that could not be accessed by an ATM card. The new technology changed our demands. The same principles could apply to printing. In the future, one may not even consider going to a printer unless the printer can offer on-demand services.

Another example is radio, which thrived at one time and was once the communication medium of choice. When TV was introduced, however, it changed the way we used the radio. Most people don't sit around the radio after they get home from work — they turn on the TV set. The way we listen to radio or consume it is in our cars or at work. The new technology changed our consumption demands.

On-demand printing will change consumer demand and consumption. This means that consumers may start to request shorter runs. Therefore, it is projected that short-run work will increase at the expense of longer runs. The table below outlines these predictions.

|      | Page Count | 5,000 Copies or Less | 5,001 Copies or More |
|------|------------|----------------------|----------------------|
| 1993 | 10 pages or under | 26.5% | 23.5% |
|      | 11 pages or more  | 38.0% | 12.0% |
| 1998 | 10 pages or under | 34.5% | 19.5% |
|      | 11 pages or more  | 34.0% | 12.0% |

## Turnaround

Today, half of all print jobs are delivered within one week. Within the decade, over 80% of all jobs will be delivered in that timeframe:

|                  | 1995 | 2005 |
|------------------|------|------|
| Less than 1 day  | —    | 14%  |
| 1 day            | 1%   | 17%  |
| Next day         | 1%   | 10%  |
| 2 days           | 21%  | 22%  |
| Within 1 week    | 28%  | 24%  |
| Within 2 weeks   | 36%  | 11%  |
| Within 1 month   | 7%   | 1%   |
| More than 1 month| 6%   | 1%   |

*Source: GAMA*

## Print Buyer Perceptions

Much of what printers sell depends upon what buyers think they are buying. The perception of the buyer is the key to selling printing. The following is a recent survey of print buyers' priorities regarding the feasibility of on-demand printing:

| | |
|---|---|
| Cost savings over offset | 92% |
| No or minimal inventory | 89% |
| Fast turnaround | 83% |
| Variability or personalization | 71% |

Quick turnaround, however, comes at a premium price. The numbers indicated in the table below represent the average percentages print buyers are willing to pay for on-demand services:

| | |
|---|---|
| Fast turnaround (<2 days) | 5% |
| No or minimal inventory | 11% |
| Variability or personalization | 19% |
| Cost savings over offset | −12% |

*Source: GAMA*

According to the same study by GAMA, buyers would pay 8% more to upgrade from one to two colors, and 11% more to upgrade to one process color, on average.

## Where Will the Work Come From?

The new processes of on-demand are projected to "rob" market share from older processes. This is typical where a technology shift is involved. Most of the work that will become short run is currently long run. This may have quite a future impact on run size.

| | |
|---|---|
| Long runs converted to short runs | 29% |
| Traditional print to digital press | 24% |
| New work | 21% |
| Monochrome converted to process | 18% |
| Color copier to digital press | 8% |

## Who Will the Users Be?

On-demand technology will be applied by those firms who are involved with actual print reproduction as opposed to copy centers. In round numbers, they are firms that currently possess printing presses or are involved with prepress operations:

| | |
|---|---|
| Commercial, book, and periodical printers | 26,000 |
| In-plant printers | 15,000 |
| Quick printers and copy shops | 14,000 |
| All other printers | 9,000 |
| Prepress services | 5,000 |
| Newspapers (with presses) | 4,000 |
| **Total** | **73,000** |

The present players in printing have the advantage. They have:
- The ability to deal with digital files
- The sales and service organization
- The manufacturing infrastructure
- Experience with reproduced products
- Experience with finishing
- The ability to offer multiple related services
- Knowledge of the printing processes

## Summary

When existing research is examined and interpreted, it becomes clear that on-demand, digital, and customized printing and publishing are growing in importance.

In the next chapter, we will discuss the economics that make these new technologies an unrelenting force, and the applications that have already demonstrated their advantages.

# Chapter 3
# Economics of On-Demand Printing

There has always been a demand for shorter printing runs, but various hurdles impeded the successful delivery of this service. The traditional methods of creating printed pieces are more expensive because of the associated prepress, press, and post-press costs. Prepress costs include the hardware (computers) and software (page layout programs) as well as price of the scanners, output devices, film, processing chemistry, and plates. Press costs include makeready costs and usually long printing runs due to price breaks for longer runs. Postpress costs include the time and cost of collating, binding, and finishing.

These costs are significantly lower in on-demand printing due to shorter makeready time, lack of prepress work, and in-line binding and finishing.

Some of these steps remain important regardless of printing technology, such as the coordination of design, typesetting, and page layout. Some steps are associated with traditional offset printing such as photography, additional film work, stripping, manual imposition, proofing, platemaking, and press start-up, but are later consolidated with improved workflows. The desktop workflow, for example, usually uses scanning instead of photography, electronic imposition instead of manual imposition, elimination of stripping by using imposition programs, PostScript, and imagesetters, and/or platesetters.

## The Economics of Long Runs

Traditional printing costs become more cost-effective, in terms of per-unit costs, as run lengths increase. In contrast, electrophotographic or photocopying technologies have fixed costs that remain the same regardless of run length. Copying usually does not require additional film work, stripping, manual imposition, proofing, platemaking, or press start-up. These factors, combined with the low cost of the equipment and toner, allow copying to be cost-effective for very short runs.

As a result, copying results in fixed costs that do not change as run length increases. In contrast, offset, and other forms of printing, result in variable costs. Thus, the per-unit cost of copying does not change as the run length changes, but the per-unit cost of printing is reduced as run length increases.

New technologies, such as computer-to-plate and automated presses, are applying techniques that allow a press to perform makeready in record time and are thus able

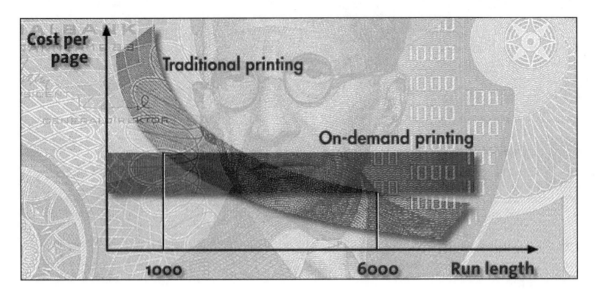

*Courtesy of Agfa, a division of Bayer USA*

to produce shorter runs. On-demand concepts, therefore, do not exclude the use of traditional printing presses.

As you can see in the illustration above, on-demand printing, utilizing electrophotographic techniques, is clearly advantageous up to 1,000 pages. In the range from 1,000 to 6,000, however, there is overlap, and the technology that is less expensive depends on the specific job.

Five thousand single-sided sheets, for example, might be completed faster and more economically on a traditional press while 2,500 double-sided sheets might be better off on a Xeikon DCP-1 or Agfa Chromapress.

## Advantage for the Buyer

The benefits of demand printing for print buyers are substantial — it allows companies to introduce products to the market faster, make last-minute revisions, and reduce or eliminate costly warehousing. And, as we will discuss in a later chapter, digital printing allows for the creation of personalized and custom-printed products for use in targeted marketing campaigns.

Another advantage of on-demand is that it makes shorter runs more economical. Remember, according to our numbers, approximately 56% of commercial, book, and office printing today is done between 500 to 5,000 impressions. And according to the CAPV data, by the year 2000 approximately 50% of all work will be short run. This means that people will stop buying longer runs than they need in an attempt to reduce their per-page or per-unit cost.

Although buying in larger amounts is great in theory, two problems often occur that prevent the customer from realizing the savings. One problem is that the customer will never use the balance of the run or when the customer attempts to use the excess material, it is often outdated or needs to be improved. Either way, if the client does not use the additional copies, then the client pays a higher per-unit cost in the end.

## The Efficiencies of Digital and On-Demand Workflow

The time and cost to manufacture a printed product using a digital workflow is typically less then creating the same piece using a conventional workflow. Here's an example of the workflow, time, and costs to prepare and print a page in color at a quantity of 500. Since there are so many possible workflows, this is merely an example:

1. *Creative development.* The creative area has been most affected by desktop technology. Most art and design professionals now create their pages with desktop computers. It is the by-product of this process that results in electronic pages. These costs should not be calculated in the final cost of the reproduction of the job.

   We repeat — none of the costs associated with the creative process are calculated in the cost of reproducing pages. For some, this is an issue. If a financially trained person calculated a return on investment (ROI) equation for a purchase, he or she would calculate the equipment, personnel, and overhead. Another way to think about this is that, in the traditional workflow, the creative personnel acquired type and art from commercial services and were charged. Those charges exist no matter how the job is reproduced.

2. *Film output.* Most production work today is done by the designer. He or she creates the pages, proofs them on low-cost laser printers, and sends them to a service bureau for printout on photographic paper or film. Costs vary significantly, but most printouts are at about $7 per page for photo paper and $19 for film (black and white). A set of four films with a film-based proof for four-color printing would probably sell for about $60.

3. *Separations and proofing.* Just as the price of output varies, the price of scanning and proofing vary as well. A further complicating factor is that alternative image acquisition options are continuously eating away at the amount of work done on high-end scanners. Therefore, digital photography, Photo CD, and desktop scanners will be used more and more regardless of the workflow or printing technology. For this example however, the cost is $60.

4. *Stripping.* Just as typographers have become extinct, so too will strippers. (Sorry, it's not our fault.) Traditionally, strippers would inspect the film for quality and pinholes, opaque the pinholes with a special brush and liquid so that light would not pass through, strip the page units onto a special sheet, and finally inspect for alignment and quality. Once again, consumables (the special sheet), labor, and equipment are the cost areas. Printing industry cost standards are presently about $60 per hour, or about $35 per page. For our purposes, the cost is approximately $70 per page.

5. *Platemaking.* Platemaking, the final step in the prepress process, is varied for the different types of printing. For the sake of simplicity, let's say that a printer uses a photographic process to expose the reversed image from the negative film separation onto a flat metal or paper plate. At this point, the image on the plate is a right-reading (you can read it). It has a chemical coating that attracts ink but not water while the nonimage area attracts water but repels ink. Platemaking typically costs about $70.

6. ***Makeready/print.*** Makeready is essentially all the work done to set up a printing press to deliver the first page. For this example, let's say that it takes one hour for both the makeready and printing at $100.

7. ***Paper.*** Paper costs have been escalating, and paper is also becoming more difficult to get. The cost of paper includes the required stock, including any waste for makeready. For this job, we'll say we used some leftover paper that the printer wanted to get rid of that was $15. (Once again, paper costs are highly variable, and should not always be calculated in a comparison of machine costs.)

To summarize traditional reproduction:

|  | Per page | Hours |
|---|---|---|
| Film output | $60 | 1.0 |
| Color separations | 60 | 0.5 |
| Stripping | 70 | 1.0 |
| Platemaking | 70 | 0.5 |
| Makeready/printing | 100 | 1.0 |
| Paper | 15 | — |
| **Total** | **$375** | **4.0** |

These costs may vary from printer to printer, and region to region, but major industry associations publish cost standards so that all printers can measure their productivity and cost-effectiveness.

As in the following example, the cost of conventional color reproduction carries high prepress and preparatory costs. This immediately negates the ability to produce short runs cost-effectively.

Alternative workflows utilizing desktop and electronic systems are attempting to bypass some of those steps. One of the advantages of the PostScript workflow is that you print four pieces of final film from a PostScript imagesetter, eliminating costs associated with stripping and platemaking.

|  | Per page | Hours |
|---|---|---|
| Film output (RIP time) | $20 | 0.5 |
| Color separations | 40 | 0.5 |
| Stripping | — | — |
| Platemaking | — | — |
| Printing | 200 | 0.5 |
| Paper | 10 | — |
| **Total** | **$270** | **1.5** |

In this example, on-demand reduces the cost of printing by 28%, and the time by 62.5%. We believe that using a digital press workflow offers a similar set of savings to the desktop workflow, as the desktop workflow provides for the conventional workflow. If we used a digital press, several steps would be eliminated. With digital printing there is no stripping, platemaking, or makeready. As a result, digital printing can result in a savings of about 30% in dollars and 60% in time.

## The Short-Run Pricing Paradox

These numbers showing cost and time savings might lead you to believe that anything you print digitally or on-demand will be faster and cheaper. Of course, it's not that simple. Remember, this was a very simple example; there was no trapping, imposition, or finishing. Many factors such as these determine how much money a company charges for its digital printing services. One factor is the cost of manufacturing.

The first factor to calculate in reaching a cost of manufacturing is the price of the equipment. Next, the maximum number of pages created per time unit must be calculated, which will help to establish a per-page price which will ultimately pay for the equipment in three years or less.

When comparing the cost of on-demand to the cost of traditional printing, the cost of collating, binding, and shipping must be taken into account because these costs may be eliminated. For example, many on-demand presses produce small books which are already collated and bound. This saves the additional cost of binding and finishing in traditional offset printing. In addition, if pages are created digitally, and the files can be sent via modem to other locations for printing, shipping charges can be reduced as well.

When calculating the costs associated with demand printing, be sure to take into account the cost of consumables and downtime. The cost of consumables for some devices may be as much as 75¢ per page, which may increase the price. Downtime due to breakdown or maintenance is also a concern as you may be able to operate an on-demand press for only six hours in an eight-hour shift.

Lack of competition or excessive equipment downtime can drive prices down to rock bottom. In a recent survey of twelve Indigo E-Print 1000 users, we requested prices on 4/0 and 4/4, 11×17-in. prints in quantities of 250, 500, 1,000, and 5,000 from QuarkXPress files. Six of the twelve companies faxed back prices almost instantly. The others are still trickling in. (Evidently they print faster than they quote.) The results were amazing:

| Company | 4/0 250 | 4/0 1000 | 4/4 250 | 4/4 500 |
|---------|---------|----------|---------|---------|
| A | $3.50 | $3.50 | $7.00 | $7.00 |
| B | 2.14 | 1.24 | 4.00 | 2.44 |
| C | 2.76 | 1.50 | 4.20 | 2.70 |
| D | 2.01 | 1.43 | 3.09 | 2.04 |
| E | 2.90 | 1.71 | 5.40 | 3.19 |
| F | 2.63 | 0.53 | 3.68 | 0.60 |

Company A had an interesting approach: a flat rate regardless of quantity. It was the highest of all of the rates. Companies B through E fell into a definable range.

Company F would not quote beyond a quantity of 250 on the Indigo. For all other quantities, they quoted with offset printing on lithographic printing presses. As indicated, the rates were extremely competitive.

This is a brilliant marketing strategy. The Indigo gets the customer in the door, and the company, which has installed the latest prepress and printing equipment, gets an opportunity to promote its message to the market.

Short-run color printing is a moving target. The printing process continues to apply more and more cost-cutting and productivity-enhancing technology. Where in the past short run meant 5,000 because of the makeready process, it may now be closer to 3,000, or even as low as 2,000. Most users of Indigo E-Print 1000 and Xeikon DCP-1 presses have average runs of about 300 — and based on their costs this is probably appropriate.

Company F would not generate the volume in short-run color without the lure of the Indigo. We will see more of this approach as users discover that each reproduction technology has an applicable niche based on run length.

Because technology gives suppliers the ability to do things faster or inexpensively does not mean that the selling price will be less or the turnaround time faster. Factors other than technology impact the cycle time and charges.

# Chapter 4

# Database Marketing's Role

Another factor motivating interest in on-demand and digital printing is an emerging trend from marketers and retailers to better target their advertising dollars. Those who belong to organizations such as the Direct Marketing Association or believe articles in popular business publications (i.e., *Business Week, Forbes, Fortune*) talk more and more about the evolution of new marketing trends.

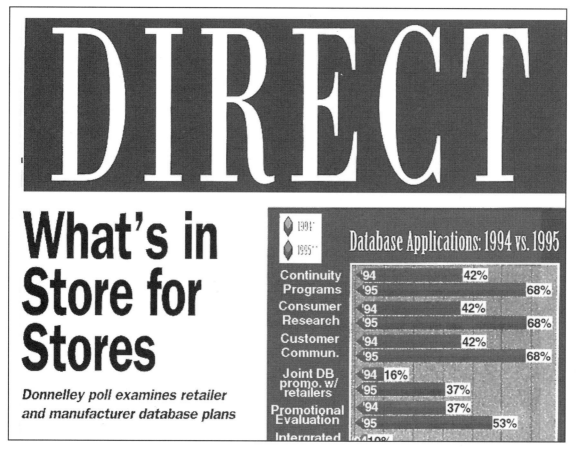

Cover story of *Direct Magazine*, 8/95, on the use of database marketing.

In the first trend, called "mass marketing," advertisers sent the same message to a "vast, undifferentiated body of consumers who received identical, mass-produced products and messages." This is often jokingly referred to by the Henry Ford slogan, "you can have any color car you want as long as it's black."

The second, more current trend, is to divide consumers into smaller groups with common demographics. Referred to as "market segmentation," the strategy was to divide anonymous consumers into smaller groups with common demographic characteristics to predict buying intentions. Using this approach, for instance, a 34-year-old, married male, with two children, a house, and earning $50,000 annually would be a candidate for a mini-van advertisement because they are the segment of the population considered most likely to buy.

This hypothesis, that demographics predicts purchase intentions, is under close scrutiny. The evidence suggests that demographics do not accurately or frequently predict purchase intentions, but are more accurately predicted by previous buying patterns.

This latest trend, "database marketing," uses the philosophy that a more accurate indication of purchase intentions is based on what has been purchased in the past. This technology has become possible due to new technology that enables marketers and retailers to pinpoint smaller and smaller niches. These faster, less-expensive computers enable marketers to zero in on small niches of the population, ultimately aiming for the smallest consumer segment of all—the individual.

Marketers now closely monitor what we buy. This is evident in any supermarket that scans your purchases. They track what we buy, when we buy it.

Although relatively new, organizations are now saving and sorting through this information, and more and more marketers are building databases that enable them to discern who their customers are, what they buy, how often they buy, and what they want.

For years, members of the Direct Marketing Association have advocated these strategies, and catalogers, record and book clubs, and credit card companies have successfully used these strategies to market their products and services.

These strategies are now moving into the mainstream. Today, companies ranging from packaged-goods to auto makers realize that in the fragmented, highly competitive marketplace of the '90s, nothing is more powerful than knowledge about customers' individual practices and preferences.

If this marketing trend continues, traditional print production would be placed at a serious competitive disadvantage. In traditional print production, when 50,000 impressions are printed, they are impressions of the same exact piece. There is no unique message sent to an individual. Other forms of media, however, such as online services, can easily customize messages.

## Customizing Traditional Print

There are techniques available today that can customize traditional offset print. The two best examples are inkjet and Selectronic™ binding. Most of us have seen inkjet personalization. It has the ability to stop us dead in our tracks. Thumbing through one of your magazines, you notice something that catches your eye — your name printed on an ad saying "John Doe, this product is for you."

Selectronic™ binding is less obvious. A large automotive manufacturing company may buy a page in a popular magazine and advertise a convertible in the Florida version, and a sport-utility vehicle in the Denver version based on demographic information. Although inkjet and Selectronic™ binding are both available, they are only used by a small portion of advertisers. The first publicized demonstrations of inkjet personalization technology occurred at the Print 91 trade show in Chicago. Selectronic™ binding technology has been around for more than 15 years. Some of the earliest reports of the technology discuss how Donnelley used it in early in the 1980s. Today other printers such as World Color Press, Quad/Graphics, and Perry Printing have these capabilities.

If database marketing and personalization are becoming more important, it is difficult to understand why inkjet and Selectronic™ binding have not grown faster. According to *Folio* magazine (Nov. 1, 1993) there are two reasons for the slow growth of Selectronic™ binding.

One problem with database-generated Selectronic™ binding is that in most fulfillment systems it tends to slow down the process. For example, if you want the bindery to insert certain pages into one magazine and withhold them from another, your fulfillment tape must contain complex instructions, which will ultimately slow down the equipment.

In addition, most fulfillment houses cannot process the load of data needed for individual customization and therefore can't perform complex segmentations on their own. Although marketing databases do have that capability, they usually are not updated frequently enough to create weekly or even monthly tapes. Even if they are, there is often a time-lag between the transmission from the database to the fulfillment file.

The most important factor, however, for deterring the acceptance of Selectronic™ binding is the fear that the technology will increase production costs. For those who are not leery of the price, there is the additional challenge of convincing advertisers to pay more. The publishers must decide if the additional costs can be offset by increased revenues from advertising and/or circulation.

For the most part, databases and publishers are not equipped to handle Selectronic™ binding. But, publishers are giving more attention to the technology due to demands from advertisers and the possibility of more effective promotions.

## Customized On-demand Print

Utilizing on-demand printing, direct digital color printing, and database marketing, retailers can send consumers personalized advertising which targets the products or product categories they sell. Instead of getting a Sunday newspaper insert in which 90% of the products are not of interest to you, you will get an insert in which 90% of the products or product categories are of interest.

In some ways, it could be the end of true junk mail. Lawn mower ads would go only to homeowners, and diaper ads would only go to parents. Combining on-demand printing or direct digital color printing would allow retailers to better target their products.

Finally, an argument could be made that the ability to customize printed products might be the only way for print to remain competitive with emerging media technology. Remember, though, that most printing is subsidized by advertising. Without advertising, newspapers and magazines would not exist.

Since advertisers have a choice of where to spend their dollars, they typically spend them in focused or narrow markets. When 500 television channels are avail-

The illustration at the left is a black-and-white version of a color document that used variable information to customize brochures for sales representatives selling computer equipment. On the far left in the illustration, you can see the area that the salesperson's name would appear.

*Courtesy of Agfa, a division of Bayer USA*

able, retailers will have the ability to focus their advertising dollars to specific audiences more then they can today with print advertising.

On-demand printing, direct digital color printing, and database marketing, however, will allow print to be a more viable method of customizing and distributing advertising and will keep the printing industry competitive in today's cutthroat market.

## The Future

The amount of purchases made through database marketing is minuscule in comparison to the amount of money spent in general retailing. According to *Fortune* magazine (April 18, 1994), Americans spent approximately $60 billion in 1993 through catalogs, TV shopping channels, and other direct-marketing alternatives. That only accounted for 2.8% of the nation's total annual expenditure of $2.1 trillion per year, however. Included in this figure are purchases made in supermarkets, mall outlets, car dealerships, department stores, warehouse clubs, boutiques, and other sources.

Marketers and merchants expect the amount of money spent through conventional retailing channels to remain steady or contract slightly as we approach the year 2000. Using new database marketing techniques, especially the customization and personalization features of the digital presses, database marketing should increase to 15% of total sales. With annual revenues of well over $300 billion, that would make database marketing one of the world's largest industries.

This book, *The One to One Future,* is a good source for information about selling to individuals or the "market of one."

# Chapter 5
# Other Forces of Change

Like other communication industries today, the publishing and printing industry is undergoing dramatic changes. The emergence of new technologies such as desktop computers and PostScript output devices combined with the increasing changes in print production, and environmental concerns, have contributed to a difficult period of transition.

Print customers have also heavily invested in desktop publishing equipment in order to send electronic files to their printers. Commercial printers have spent the last few years restructuring their businesses to meet the changing needs of their customers as a result. Today, printers receive more than 50% of their jobs in electronic form.

Other forces of change such as paper costs, postal rates, CD-ROMs, and a national preoccupation with online services have prompted print customers to investigate alternative production and distribution options. The methods used to prepare, create, and distribute traditional printed products are currently under fire.

A traditional printed product is one that utilizes plates, film, prepress, printing, finishing, warehouse storage, and ground-based transportation. Conversely, nontraditional or alternative products are those not requiring plates, film, long press-runs, warehouse storage, and ground-based transportation such as CD-ROM, online products, and materials created by on-demand, digital, and customized print.

New forces competing with traditional print such as online services, CD-ROM publishing, and interactive TV have advantages as alternative products or alternative media. Once we define the advantages and disadvantages of traditional print and alternative products, we can figure out how print can compete in the new communications world order.

With advantages over traditional print, such as customization, lower cost, and timeliness, digital and custom printing could be considered an alternative media. On-demand, digital, and custom printing and publishing, therefore, are considered alternative or new media, and as new media products they are in the same category as online services, CD-ROM publishing, and interactive TV.

In assessing the advantages of different types of media we could discuss interactivity, attractiveness to advertisers, search and retrieval advantages, manufacturing cost, target effectiveness, and method of delivery — all relevant considerations in making decision regarding which media will best suit your needs.

## The Interactivity Advantage

According to media guru Jack Powers there are six advantages of interactive publishing: customization, timeliness, comprehensiveness, searchability, transaction, and economy. He says that few other types of publishing offer all of these benefits.

While a fax publication is timely and customizable, it is not comprehensive. A CD-ROM is economical and comprehensive, but it is also quickly out of date. It can also be difficult to complete transactions via CD-ROM because CD-ROM is not online. Recently, however, CD-ROM catalogs have been distributed with links to online services (i.e., America Online) so that orders can be placed directly. While these electronic catalogs offer timeliness and immediate transactions, they may not be customized to certain preferences as other media have the ability to do.

According to Stephen Manes *(NY Times,* 12/13/94) the importance of interactivity is hardly new. It can be found in all facets of life. Books divided into chapters allow browsing. Experimental plays and films sometimes let audiences influence the action. Perhaps the most popular interactive medium is the video game, which was invented in the early 1960s. But in recent years interactive works have latched onto the delivery medium of CD-ROM and online services.

Photo CD combines CD-ROM and TV technologies with an interactive display.

If advertisers influence the success of certain media and interaction enhances advertising, then you have to wonder if advertisers will consider interactive new media (CD-ROM, online) a better buy than noninteractive media (newspapers, magazines).

Some say yes. A Sega/Nintendo video game, for example, offers avant garde advertisers potential levels of customer interaction that no traditional 30-second TV spot could dream of. What better way to burn a sales message directly onto the brain of an impressionable teenager than to craft an exciting, high-speed video game around it?

Interactivity is a continuum, and all communication methods offer some level of interactivity. A newspaper with a table of contents allows you to "look at a glance" and decide what you want to read. Compared to a novel, a newspaper is much more interactive. CD-ROM products offer more interactivity than most printed products. The ability to "mouse around" allows you to create your own path through a CD-ROM. Printed products offer certain interactivity, but far less than a CD-ROM.

A few years ago an article in *New Media* magazine said, "Last year, Chrysler Corp. sold about 6,000 marketing disks for $6.95 each to show off its Jeep and Eagle models." The detailed presentation allowed viewers to preview several of the latest models on their computer screen. "For years, software companies have been sending out demo disks to prospective customers," says John Clark, a consultant with Mar-Tec International. "This disk-based marketing approach has crossed over to include any industry that can produce a printed brochure."

## Online Interactivity — Advantage
Unless you've just arrived on this planet from some far off galaxy, by now you must have seen, heard, or "surfed" a commercial online service (America Online, Compu-Serve, Prodigy, etc.) or the Internet. The emerging vision of an online future in "cyberspace" has been the cover story of almost every popular magazine.

Everyone from politicians such as Al Gore to the billionaire chairman of Microsoft, Bill Gates, have painted a detailed picture of future life online. Many people, especially academics, already work, shop, chat, and educate each other online. The future vision includes a new online realm that will put every conceivable form of information—from a newspaper or magazine to a digitized, interactive movie—online and available to anyone who has a computer and a modem.

Millions of people collaborate daily on computer networks or log on to commercial online services to surf the Internet. Until recently, a select realm of universities and government agencies were privy to Internet access. New Internet service providers and new web-browsing software have emerged, making it possible for anyone to access the Internet. This big breakthrough began in 1993 with the creation of a subsection of the Internet called the World Wide Web (WWW).

The WWW has two advantages: it is extremely user-friendly and is connected to other sites around the globe. Schools, government agencies, and businesses can build what is called a "home page." The home page can be thought of as a store front in a shopping mall. It uses a graphical user interface (GUI) so you can see "at a glance" what is on any home page and you can then decide if you want to see it or go somewhere else.

The other advantage is the linking or the "hyperlinking" ability. Unlike conventional online services such as America Online, Compuserve, and Prodigy, the Internet is made up of thousands of computers. With a web browser, which is software for the Internet, when you come to a word or sentence that is underlined or in a different color, you simply "click" on that area with your cursor or mouse, and you are instantly launched or sent to another computer that discusses that word or sentence in more depth.

While the Web is relatively new, it's the fastest-growing segment of the Internet. Web browsers such as Internet Explorer and Netscape allow you to interact with the Web sites and other Internet resources with point-and-click ease. In one stroke, the Web makes the Internet much easier to use and gives you the graphical tools to set yourself up as an information publisher.

All of this Internet activity has created a new communication medium. People looking for jobs are posting multimedia resumes to the Web, companies testing new marketing strategies are creating electronic storefronts, and even rock-and-roll bands like the Rolling Stones have a home page. Software, fonts, graphic images, and shareware can be downloaded from all over the world.

Underneath all of the publicity, the Web is nothing more than a set of hyperlinked elements that conform to a standard known as the Hyper Text Markup Language (HTML). HTML is a subset of the Standard Generalized Markup Language (SGML), a standard for cross-platform publishing. SGML allows you to translate documents to other platforms and other media, such as CD-ROM. HTML, on the other hand, is not limited to text but can include audio, video, text, graphics, and tables, as well as links to resources like e-mail, UseNet news, and other Web sites.

Earlier, we posed the hypothesis that if print production is subsidized by advertising, then advertisers will emphasize the medium that sells the best. Think about this — are online services a good vehicle for advertising? Better yet, are online services a good vehicle for selling?

## Interactive TV

Pretend it's the year 2005. You're watching the movie *Star Wars XXIII* on a new TV-on-demand service, and it's working great. It's part of your phone company's two-way

interactive-TV system, which is designed to demonstrate how much more time we can spend in the safety of our houses without human interaction. Using the system, you can order movies, television, and even pizza delivery from a menu of listings.

Currently, tests are being conducted all over the country to determine the viability of interactive TV. The key component is the so-called "set-top" box — the device that sits on top of TVs and unscrambles the cable signal. Several companies have announced plans to manufacture set-top boxes. General Instrument, for instance, has announced plans to introduce a set-top box made of an Intel 386 microprocessor and an operating system designed by Microsoft. Other announcements have been made by Scientific-Atlanta, Motorola, and Kaleida (a joint venture of IBM and Apple) to roll out set-tops of their own.

*Courtesy Eastman Kodak Co.*

It is uncertain if interactive television will be a cable- or phone-based service. The first method of distribution will most likely be through a cable company that will buy and lease them. But eventually each household will be able to buy a set-top in a consumer electronics shop. Silicon Graphics, 3DO, HP, and Toshiba are among those planning to retail set-tops. These boxes are expected to offer 3-D graphics, access to specialized interactive services, and other advanced features and will sell for approximately $500–700.

Unlike standard cable boxes, which do little more than decode the scrambled cable feed, these new set-top boxes will come with built-in computers to receive, decompress, and allow users to play interactive games. The built-in software will be user-friendly enough to allow anyone to successfully interact.

Microsoft's set-top interface will be "based on Windows technology," says Karl Buhl, spokesperson for Microsoft's advanced consumer technology group, "but it won't bear much resemblance to Windows. Instead of a screen with lots of buttons, you'll see very simple graphic overlays" on top of existing programs. "Users will be able to select programs by category (such as news or comedy shows), 'rent' movies, buy that new backyard grill, or play Jeopardy — all with a few clicks of the remote."

According to Bill Gates, founder of Microsoft Corporation, the future of shopping is online. Not online as we know it today, but the interactive TV of the future. Gates says, "You're watching Seinfeld on TV, and you like the jacket he's wearing. You click on it with your remote control. The show pauses and a Windows-style drop-down menu appears at the top of the screen, asking if you want to buy it. You click on 'yes.' The next menu offers you a choice of colors; you click on black. Another menu lists your credit cards, asking which one you'll use for this purchase. Click on MasterCard or whatever. Which address should the jacket go to, your office or your home or your cabin? Click on one address and you're done — the menus disappear and Seinfeld picks up where it left off."

Today, the process of home shopping is a three-part cable-TV, 800-number, and credit-card number transaction. But this transaction is poised to move to a higher level of interactivity. In the future interactive-TV programs and in-store kiosks known as "electronic mirrors" with holographic images will enable us to see what clothes look like on our bodies without actually trying anything on. Before building or driving to a house, a computer catalog will someday include virtual-reality "tours" of each room in a house.

## Demographics

Often overlooked when someone is trying to sell the idea that you should create your own home page and advertise your services because of the millions of interactions that occur online is the fact that most of them are e-mail. Another often-neglected fact is that all that e-mail is plain text that can be intercepted. Even if your computer and addressee are secure, the nature of the Net is to route your messages through other machines, which may be accessible. Passwords do not guarantee protection because they can be intercepted and decoded.

What are the demographics of online services? According to most of the existing data, the majority of the people online are young men. Certainly, the Internet is growing and attracting a wider audience, but we will have to verify its demographics in the future. Print, on the other hand, has a track record and its demographics are trackable.

| Service | # people | % male | % female |
|---|---|---|---|
| Internet* | 25 million | 87% | 13% |
| CompuServe† | 3.2 million | 90% | 10% |
| America Online† | 3.5 million | 85% | 15% |
| Prodigy† | 1.5 million | 62% | 38% |

*MIT census of UseNet, average age 31 years old
†1995 numbers

Some people are trying to reverse this trend. For example, a former "Well" user (very popular on-line site), built a system on the East Coast and offered free accounts to women, hoping they would provide a "civilizing force" to counterbalance the Internet's male-dominated presence. It has been successful, and plans are in order to build similar services in six U.S. cities, including Boston, Minneapolis, and Los Angeles.

## Advantages of Search and Retrieval

It seems that whenever we present a seminar on new media the publishers start wondering if they should start putting their publications on CD-ROMs. To answer that question, we should look at the advantages of CD-ROM and the publications that have fared best.

The greatest advantage of CD-ROM is that you can put a lot of information on them. CD-ROMs can store 650 MB without compression and 1.3 GB with compression. This translates to about 200,000 pages of text alone or five file cabinets of paper. A single CD-ROM can bring hundreds of thousands of pages of information to your screen, from entire encyclopedias to the complete phone directory of the United States.

But raw storage is practically useless without extensive searchability. This advantage is most clearly illustrated with reference materials such as a dictionary, encyclopedia, thesaurus, etc. The reason reference materials are selling well on CD-ROMs is because you can search the entire encyclopedia in less then one minute.

The combination of search and retrieval and other new media advantages of CD-ROMs have already vanquished one traditional market in the book publishing segment — the $700 million market for reference books like dictionaries and encyclopedias. Now, sales of CD-ROMs exceed those of the printed versions.

This trend is not new; in fact, it is several years old. According to an article in *Fortune* magazine published Oct. 19, 1992, reference works on CD-ROM were outselling their weighty counterparts an estimated 150,000 units to 100,000 units. Grolier's 21-volume encyclopedia on disk, weighs a paltry 0.6 oz., compared to the traditional printed version that weighs 62 lb.

The next target will most likely be the $9 billion market for textbooks and professional books. According to Nader Darehshori, chairman of Houghton Mifflin: "Within 20 years, most study materials will be computer-based, not printed."

## Alternative Media: CD-ROMs

When Marconi invented the wireless telegraph in the 1800s, he never would have guessed how his invention would launch the wireless broadcasting industry nearly 20 years later. The result today is that radios and TVs are the most prevalent household appliances.

We can gain great insights by studying the precursors to modern technology and the resulting revolutions. You might wonder if the CD-ROM is the enabling technology that will propel us into the future technologies, such as interactive virtual reality.

But returning to this time and space, what are the advantages of CD-ROMs? We have already discussed storage and search-and-retrieval advantages. Other advantages of CD-ROMs include merging a variety of sources such as type, film, videotape, and audio to create a new type of experience that combines the best features of words, sounds, and images. What applications benefit from this combination of features?

We've already seen how reference materials have been affected. Educational materials are not far behind. The ability to click on a button and hear an inspiring speech from John F. Kennedy or Martin Luther King makes the learning experience from a CD-ROM much richer than from a book in print.

*Courtesy Eastman Kodak Co.*

Do CD-ROMs offer advertisers advantages over printed products? Some people say yes. Marketers, retailers, and mail-order companies are intrigued by this interactive marketing tool. In New York City, Tiffany & Co. participated in a pilot project sponsored by Apple Computer that produced a shopping disk called En Passant that targeted 30,000 home shoppers. Participants received a free CD that included two dozen catalogs from companies such as Tiffany, L.L. Bean, Lands' End, the Pottery Barn, and Apple.

Tiffany & Co. senior vice president of marketing Diana Lyne said the En Passant project was a low-cost, low-risk way to explore interactive shopping. But she was disappointed because it didn't create a lot of sales. The next attempt will target a different audience: business-to-business sales. "Corporations are increasingly likely to have CD-ROM drives in their PCs, so they will be able to use our catalog program," says Lyne. "To be able to quickly find a travel-related gift or a popular gift for under $100 is very powerful."

According to *Direct* magazine (12/22/94) almost two-thirds, or 63.6%, of the 565 direct marketers polled will boost spending on marketing via CD-ROM next year. More than 61% said they will increase the amount of money spent on online marketing through such networks as Prodigy, America Online, and CompuServe.

Another advantage of CD-ROM products is it allows mail-order marketers to reach their targets more effectively. In an article in *CD-ROM World* (Sept. 1994), Tony Burns, national marketing manager at Environmental Systems Research Institute in Redlands, Calif., said the technology is "a key that can unlock the whole world."

He said CDs offer an inexpensive and flexible alternative to the hit-and-miss approach of buying labels for one-time use. Instead of paying thousands of dollars for address lists, companies can spend one-tenth of that on a disc and reap unlimited access.

One of the reasons CD-ROM direct mail marketing has become significant competition for printed products has nothing to do with interactivity or sound and video but because of manufacturing costs. A recent development that has made printed direct mail uncompetitive is paper and postal costs.

## Manufacturing Costs: Paper

Another factor that makes alternative media attractive is paper costs. The price of basic uncoated white office paper has jumped on three occasions since August 1994, to nearly $700 a ton. Coated-paper supplies also became tight during the last quarter of 1994. The tremendous postal rate hikes of 1995 prompted catalogers to produce more catalogs, in an attempt to better-target their clients before the price increase.

As long-term contracts come up for renewal, buyers face hikes of over 25%. "Demand is stronger and stronger," says L. Scott Barnard, executive vice-president of Champion International Corp.

According to an article in *Business Week* (November 21, 1994) entitled "Suddenly Paper Is on a Burn," they say "Today, prices of some grades of paper are jumping by 25% in a month, and there's panic buying — newspaper and magazine publishers, small print shops, and manufacturers are stocking up before prices rise further."

*Business Week* concludes that much of the shortage is due to the fact that paper buyers, who once counted on instant delivery, are trying to build inventory. "We're seeing a change in inventory policy as well as a real acceleration in demand," says Dean Witter Reynolds Inc. analyst Evadna Lynn. "Excess capacity and price wars have kept down prices of toilet paper and diapers, but rising pulp costs may push up tissue prices next."

Corporate earnings may soon reflect all this. Third-quarter paper profits remained modest, but far bigger gains are expected for the fourth quarter and in 1995. "We're looking at two to three years of guaranteed tight markets, as long as we don't go into recession," says John Maine, a paper analyst with Resource Information Systems Inc., a Bedford (Mass.) consultancy.

## Manufacturing Costs: Mailing

Do you remember how the pile of stuff in your mailbox grew higher than normal around Christmas time 1994? We certainly did. We are aware of the ebb and flow of "stuff" or catalogs in our mailboxes in certain times. If this seemed greater than expected, it was. That's because with postal rates set to increase 14% in January 1995, many catalogers in the $57.4 billion catalog industry increased mailings 10–20%. After the postal rate increase, the catalogers pared down their mailing lists.

As third-class postal rates have risen from 8¢ per piece in 1975 to 25¢ or more today, mailing costs have skyrocketed to $1 or more per piece. According to the postal service, increased postage rates will add almost $4.7 billion in additional revenue this year.

Business marketers use a number of strategies to offset the higher rates. Here is a list by Markus Allen, publisher of MailGram (a direct-marketing publication based in Newtown Square, Pa.), that we downloaded from America Online.

- **Cleaning the mailing list.** Using good database software, mailing service bureaus can purge duplicate records, correct misspelled addresses, and insert proper mailing information for a relatively small fee ($7–9 per 1,000 records). This ensures that every mailed piece gets delivered.
- **Be selective with your customer mailing list.** Mail to "I'm interested" buyers rather than "maybe" prospects.
- **Use PostNet barcodes.** Reward yourself with big postage discounts by pre-bar-coding your mail. Those short and long lines can save you a small bundle.
- **Ask suppliers for a discount.** Save yourself from 2% to 7% by asking for early-pay discounts. Also ask current vendors to sharpen their pencil —"remember, if you don't ask, you don't get."
- **Don't pay sales tax on printing.** Check with your accountant. You can save up to 9% with this little known fact.

## Consequences of Increased Postal Rates

There are several consequences of the increase in postal rates. For one, it makes CD-ROM catalogs and targeted database marketing more attractive. That's good news to those creating CD-ROM catalogs and bad news for anyone creating traditional catalogs. One of the advantages of CD-ROM publications is size and weight. Where a large catalog can weight 10 lb. or more, a CD-ROM with a jewel case weighs only a few ounces.

Another consequence is that it has forced publishers to reevaluate their print production and distribution methods. Many are responding by cutting trim sizes and moving to lower basis weights of paper. According to *Folio: First Day* (2/15/95), Weider Publication's *Shape* and *Men's Fitness,* 54 of Cahners Publishing's 91 titles, and K-III Magazine's *Premiere, New York,* and *New Woman* are trimming down, while Times Mirror's *Outdoors* is moving to a lower basis weight.

Others are finding themselves frustrated by a tight supply of lighter-weight paper and waiting lines for short-cutoff machines. A trim-size reduction will have an immediate impact on postage costs, but major savings will be obtained only if the publications are printed on short-cutoff machines. With perhaps a dozen of the $12 million machines in use, a backlog of aspiring publishers has been created.

Moving to a lighter paper weight is also difficult. While the shortage of supply is cutting across all paper weights, 34-lb. and below are particularly hard hit. Many publishers also made trim-size cuts in 1991 and 1992 in the heart of the recession and are uncomfortable making further cuts.

And lastly, increasing postal rates reinforces the advantages of targeted or "focused" marketing. One of the most interesting aspects of on-demand or digital printing is the ability to combine a database with an output device with the ability to print variable information and customize or personalize marketing or advertising.

## Alternative Media — Online

These advantages are not limited to CD-ROMs. Online services also offer interactivity and search-and-retrieval ability. Market projections show that the amount of money spent for online services is increasing. According to *Direct* magazine (12/22/94), more than 61% of the 565 direct marketers polled said they will increase the amount of money spent on online marketing through such networks as Prodigy, America Online, and CompuServe.

Ironically, another use of online services is for marketing printed products. For example since 1992, readers have turned to On-line BookStore (OBS) for full-text, "distributive" Internet publishing of such titles as Nelson Mandela's autobiography, *Long Walk to Freedom* (in German and English), and Floyd Kemske's novel-in-progress about corporate takeovers and vampires.

The BookFinder service uses "real fuzzy logic" to link readers to the in-print books they need, in any of 269 languages from Arabic to Zulu, for a nominal $5 fee. Book-Finder is not a huge database catalog crunching out searches on author, title, and subject on the front end, with slashed warehouse prices on the back end. Instead, BookFinder works one-on-one with readers to link them to experienced book buyers in participating bookstores around the world, people who match readers to the right books and ship them books anywhere in the world. Credit card information is transferable via an online request form, telephone, or toll-free international fax numbers.

The advantages of online marketing are not limited to commercial books. It is also becoming a force with college textbooks. Simon & Schuster Custom Publishing has created the "College On-line" interactive service. In this new electronic community, professors and students can:

- Interact on message boards and through conference rooms with others who share their interests
- Download supplemental course materials and software
- Search through information on Simon & Schuster's published textbooks and new media
- Benefit from author- and user-developed study tips
- Share information on various disciplines

Simon & Schuster Custom Publishing provides a variety of individualized publishing services to professors and students. Working with the Prentice Hall and Allyn & Bacon college textbook publishing groups, Simon & Schuster Custom Publishing offers college instructors the opportunity to create their own textbooks and course materials. Custom textbooks can contain material selected from Prentice Hall and Allyn & Bacon publications, as well as articles and readings from almost any source, professors' original material, students' essays, course syllabi, lab reports, and whatever else a course requires.

This is further evidence about the consequences of alternative media on book publishing. The evidence supported the combination of on-demand printing and customized textbook publishing is so strong that we have dedicated an entire chapter to it.

## Commercial Online Services

Most businesses have access to commercial services. Arguably, the best commercial service for business applications is CompuServe because of the comprehensive database services. One example is IQuest, a service that contains more than 800 publications and databases covering business, government, research, and news,

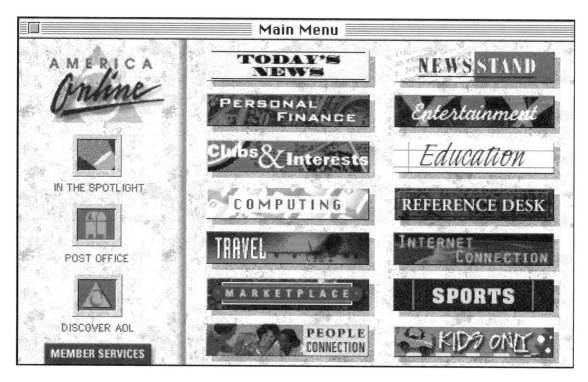

Opening screen from America Online, the fastest growing commercial service.

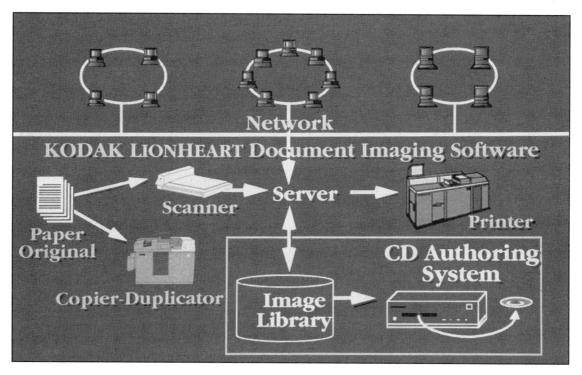

Combining on-demand printing with CD-ROM production.
*Courtesy Eastman Kodak Co.*

and looks on other online services like NewsNet and Orbit. With an easy-find feature, locating articles and reports on any subject from several different sources is quick and easy.

Unfortunately, this is one of the most expensive services that CompuServe offers. One search costs $9, there is a $2–75 surcharge depending on what databases are accessed, and retrieving an abstract costs $3. Theoretically, searching and retrieving one article could cost up to $100.

Fortunately, there are less expensive options. Depending on what kind of information you're looking for, you can often get it for a lower fee. Databases like Magazine Database Plus (Go: Database Plus) offer comprehensive lists of articles.

While the professional and expensive services on CompuServe may be too expensive, the basics provided give America Online and other online challengers a run for their money. For example, American Online with 3.5 million users is the fastest-growing service with a larger market share of home users.

Online services are becoming a valuable resource for publishers. Penguin recently published a Stephen King short story on the Internet before it appeared in print. The book entitled *Umney's Last Case* was from his latest collection, Nightmares & Dreamscapes. The entire collection is available from the On-line Book Store, a service accessible from Internet providers, as well as from ZiffNet. In addition, Time Warner's new electronic publishing unit will start a forum on CompuServe to let you preview new books and talk with authors.

## Summary
In summary there are new forces competing with traditional printing and publishing. These forces include CD-ROM publishing, online publishing, and interactive TV. These alternative media offer advantages such as interactivity, search-and-retrieval advantages, lower production and distribution costs, and marketing and advertising advantages.

On-demand printing and publishing are not traditional printing and publishing. In fact, we consider it a form of alternative media.

# Chapter 6
# Commercial Applications

Commercial printers and quick printers serve a vital role. Committed to serving the diverse printing needs, they produce a wide variety of products. Historically, quick printers or, what today is referred to as "convenience" printers, have carved out a unique niche in the printing market based on fast turnaround. However, commercial printers under competitive pressure from convenience printers are being forced to provide faster turnarounds. The problem is that the batch-oriented nature of the commercial printing business makes it difficult to remain competitive.

Printing is a batch-oriented process. Large printing presses fed by sophisticated prepress systems produce long runs. Long runs are virtually mandated in order to absorb the traditional makeready costs of the process. It is not very different from an assembly line. Commercial printing, therefore, is a manufacturing process, and like other manufacturing processes, improvements can be made either with more efficient technology or through innovative management philosophies.

Throughout history, there are many examples of technologies that have increased productivity and reduced costs. The development of phototypesetting machines driven by computers made typesetting faster than the hot metal techniques of Linotype and Monotype machines. More recently, large format imagesetters and platesetters with trapping and imposition software have eliminated time-consuming manual steps and have streamlined the workflow.

## Just in Time

Print production is a manufacturing process, and innovations to manufacturing processes, such as digital printing, benefit the entire processes. As a manufacturing process, print production could benefit from new management concepts such as "just in time" (JIT), SPC (statistical process control), and TQM (total quality management). The concept of "just-in-time" delivery of parts or supplies reduces inventories and eliminates waste for manufacturers, such as when a supplier delivers parts to the factory floor just in time for their incorporation into a product.

JIT has enabled many different manufacturers to dramatically reduce on-site storage of supplies or parts without loss. In fact, there are usually gains in productivity. Suppliers deliver goods in smaller quantities more frequently and dependably, and timed more closely to when they are needed.

Buyers of printing are seeking similar advantages. They want the ability to acquire only what they need — when they need it. This has already become evident in the book market where short-run book printing has grown significantly over the last decade. JIT philosophies are therefore affecting the purchasing of printed products.

The concept of "just in time" delivery is changing the nature of the printed product. Rather than printing long runs in order to maximize the cost idiosyncrasies of the color printing process, prudent print buyers are acquiring smaller volumes to meet immediate demands. JIT delivery not only mandates short runs, it also mandates very fast turnaround.

Major cost advantages in the JIT approaches are the reduction in warehousing costs and the ability to make changes to new runs more advantageously.

On-demand will succeed at the expense of commercial printing, we have been told. Presentations made by demand supporters point to the nature of the printing process, with its manufacturing orientation, high labor cost, commodity or low value added, materials-intensive requirement, and skill dependence.

As various analysts have researched the market, they seem to conclude that the first area where this technology will be installed will be commercial printing establishments. This is due to the fact that these establishments have the infrastructure to support the sales and service aspects of the business. Since printers might lose business because of digital printing, they will also embrace it to create new business.

We project that customers will demand shorter, more economical printing, faster and only when they need it. These advantages are possible with on-demand and digital color printing. Listed below are the applications that are the best and worst fit for on-demand, digital, and customized printing.

## Appropriate Applications for On-Demand and Digital Printing

Although we could list 500 or more applications for these technologies, instead we have listed the more popular categories. In the terms of best-fit applications, we see the following: advertising/direct mail, author reprints, books/manuals, bound galleys, brochures/booklets, catalogs, envelope/packaging, financial/legal, flyer/folder, form/coupon, invitation/menu, letterhead/stationery, newspaper, and signs/posters.

### Advertising/Direct Mail

Advertising and direct mail marketing are characterized by the heavy use of illustrations and color, with a wide variety of type styles and heavy use of graphics. Although different in their method of delivery, the goals are the same — to motivate people to buy products.

In contrast to advertisements that normally run in publications along side of editorial pages, direct mail is a stand-alone product delivered right to your mailbox. As we explain in our chapter on database marketing, direct mail can be easily customized.

On-demand digital color technology is being tested for sophisticated direct mail production. There will be an explosion of personalized promotions with personalization such as sales letters containing pictures of the sales person and the product. But could national advertising become customized?

The way we work today, the answer is "no." Today advertisements are still provided to newspapers and magazines as film or hard copy.

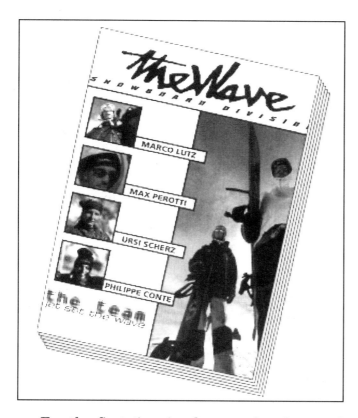

This figure depicts an interesting commercial application of on-demand printing. Herzog Imaging and Digital Printing in Switzerland created a color brochure overnight using the Agfa Chromapress system for a company attending a Snowboard championship.

To accomplish this overnight feat, Herzog received color photos from the Snowboard championship late one afternoon, and the next morning delivered 500 letter-sized leaflets before the championship ended.

*Courtesy of Agfa, a division of Bayer USA*

For the first time in almost a decade, standards committees have developed a standard for advertisement transmission called TIFF-IT. The TIFF-IT or ANSI IT 8.8 standard contains the three main components of a digital standard — image content, linework description, and high-resolution continuous-tone information content.

One of the advantages of TIFF-IT is that it allows advertisers, trade shops, publishers, and printers to continue to use their existing equipment and methods at the desktop publishing level, as well as the traditional high-end CEPS (color electronic prepress systems; e.g., Linotype-Hell, Scitex, DuPont Crosfield). This new standard will bring publications into the realm of digital printing since all parts of the publication would then be in electronic form. Digital advertising is essential for digital printing.

Once there are TIFF-IT readers and writers, publishers or printers could open up a standard advertisement and possibly customize it for certain areas or demographics.

## Author Reprints

The author of a journal article is usually entitled to 100 or so extra copies of the article to circulate to colleagues. Typically, these reprints are produced by cutting apart extra copies of the printed journal — a labor-intensive undertaking, and one that usually happens weeks after the print run.

With electronic printing, the author's copies can be created from the same PostScript files used to create the journal, and can be printed while the journal is still on the press.

## Books/Manuals

Books, characterized by high page counts, are typically black-and-white documents with a relatively consistent text format. Although black-and-white is the most popular format for books, the use of color is projected to increase. Books are one of the most appropriate applications for these technologies. (For more information, see chapter 7 about on-demand books.)

Also included in this category are manuals, technical documentation, proposals, and reports, which are the most popular electronic publishing applications. Directories are included here, with small overlap with catalogs.

One aspect of production that contributes to the cost of products in this category is collating and binding. Binding is usually adhesive (perfect binding) or sewn, although mechanical binding is also used. The dream of on-demand digital printing is that it will produce books on-demand. This may not be so much a printing problem as a binding problem.

## Bound Galleys

Initial reviews of a book are generally written before the book itself is printed. They are based on "bound galleys," which represent an intermediate stage when the content is complete, but final editing and page makeup have not occurred. Using traditional printing techniques, creating bound galleys for reviewers is extremely expensive. Although expensive, it is a necessary step if the reviews are to be published before the book hits the stores. With on-demand and digital printing, bound galleys can be created readily from an electronic manuscript at any stage.

Electronic printing is uniquely suited to the production of customized materials. If you need completely customized documents (each copy unique), you have no other choice. For example, a prominent New York brokerage firm has begun sending out customized portfolio reports to its pension-fund clients. Each fund has a different portfolio, therefore, a different report. These reports are often 40–50 pages long. They consist of a great deal of boilerplate material, but with many customer-specific variables. Electronic printing (in this case, a Kodak 1392) is the only practical way to produce documents such as these.

Semi-custom documents are also a natural for electronic printing. When referring to semi-custom, there are a number of copies (from several dozen to several hundred) of each variant, and all the variants are derived from a master database.

## Brochures/Booklets

Typically, brochures and booklets are produced from multiple-folded paper (less than 100 pages) that are grouped into sets and stapled through the fold (saddle stitched). We predict that products in this category are expected to increase in color usage as a result of the new technologies.

As we describe in the in-plant chapter, these documents are important for large companies. They are used both in internal and external communications. Also, online bindery functions will allow new electronic printers to produce completed brochures, which would increase the number of products in this category.

## Catalogs

Specialized catalogs lend themselves well to the advantages of on-demand printing and publishing, especially when combined with variable information. With electronic printing, it is possible to customize different versions.

As a result, you could select certain items appropriate for a specific region (i.e., winter coats for northern regions), a specific trade show (i.e., the Graph Expo printing show), or a specific customer type and publish a custom-tailored catalog. With the advent of high-speed, on-demand, digital printing, this type of catalog marketing will increase.

Although not as well-known, another type of catalog is the industrial catalog, which is essentially groups of product sheets. These are appropriate for digital printing using variable information.

Other products that could be included in this category are flyers, brochures, or booklets used for marketing and contain considerable process color pages. We project that more single-sheet and four-page units will be utilized once competitive costs are available.

*Courtesy of Agfa, Division of Bayer USA*

## Forms/Coupons

Fill-in forms are characterized by high usage of horizontal and vertical rules. It is difficult to generalize about the content because the formats are very erratic, except for the fact that they contain lines of text and a place for a signature.

Essentially, a form contains some amount of canned or repetitive data combined with variable or personalized data. This is a perfect application for digital printing since it has the long-term ability to change the page image on each page, where traditional printing repeats the same image on every page.

## Magazine Reprints

According to Barbara Schetter, vice president of the Digital Division of R.R. Donnelley & Sons, magazine reprints are an excellent application for digital and on-demand technology. Often, when a company orders reprints, they are used as sales tools. Since there is often an unprinted area or white space on the reprint, it is well suited for customized printing. Schetter says, "therefore, when a customer wants a reprint describing a certain piece of equipment, we could print it for them and include the local store name, local salespersons, even a coupon on it."

Another advantage is that the magazine reprints can be ready when the article hits the newsstand. Today, it may take weeks or months to receive reprints. Often the publisher, printer, or fulfillment company waits until there is enough pent-up demand to justify a print run. With digital and on-demand technology, reprints can be made faster, for less money, and customized for specific purposes such as sales.

## Newspapers

While the benefits of this technology to most other applications are fairly obvious, this category often surprises people. That is because newspapers are large-circulation publications and large in format size. Categorized as either broadsheet or tabloid (half a broadsheet), they are printed and folded in one pass through the printing press, often containing several sections and inserts.

*Courtesy of Agfa, Division of Bayer USA*

In the last few years the cry that "newspapers are dead" has been heard across the globe and has been reiterated by a recent U.S. study. Critics claim that newspapers are failing because they are unable to offer new products and services to ward off the threat of alternative media, or because newspaper readership remains "flat" or stagnant. They also emphasize how newspapers fail to target audiences with appropriate editorial or advertising opportunities.

Although leveled at newspapers, these criticisms could be made of other publications. Not all newspaper circulation is dwindling, and many are offering new services such as online services to insure their longevity.

The study "Newspapers into the 21st Century" published by GAMIS (Graphic Arts Marketing Information Service), a division of the PIA (Printing Industries of America), adds more fuel to the fire.

The study says that from 1970 to 1991 circulation in the United States remained flat at between 60 and 63 million while the total number of dailies decreased from 1,748 to 1,611. The number of readers fell below the growth of the population. In the last 20 years while the number of adults grew by 37% and the number of households increased by 46%, circulation increased by only 1.1%. And over the last 20 years the number of readers has decreased; in 1970 78% of adults read newspapers and today only 62% read them.

Due to the emerging factors that we listed in the first chapter, alternative production and distribution techniques are in beta testing by newspaper publishers.

It is important to disassociate the "death" of newspaper publishing from the decline of newspaper publishing on paper. Newspaper readership of the printed version may decline, but it is unlikely it will die. The "print" version will evolve, as the radio did with television, into more specialized versions.

When it comes to news delivery, however, newspapers that are quick to adopt alternative delivery will not only survive, but thrive. This alternative delivery could be online service, fax publishing, or on-demand printing.

Some newspapers use fax delivery. Currently, the *New York Times* and *London Times* produce a fax newspaper that is sent to certain hotels, cruise ships, and resorts. The fax newspaper is printed on laser printers and reproduced on copy machines or even printing presses. It can also be received over the World Wide Web where it is available in Adobe Acrobat form.

Many newspapers are already online. During tax season, for example, the *San Jose Mercury News* offered a fax delivery service of federal and local tax forms as well as a conference-call service between readers, a CPA, and the manager of the paper's finance department. Since July 1993, the *Chicago Tribune* has offered a service called Chicago online that offers clients a BBS for access to the printed content of every issue.

In addition, the *Tribune* has recently added a new service that allows readers to access the information that may have been cut from the articles because of space constraints. The *Washington Post* has announced plans for its own online service that, besides offering the publication in an electronic form, will also allow exchanges between readers and writers. The *Los Angeles Times, San Francisco Chronicle,* and *Washington Post* have also announced similar online services.

## More On-Demand Products

There is virtually no end to on-demand product types. Here are some other products that we believe are well-suited for the technology:

- **Envelope/packaging.** Because of the unique creation and production requirements, we believe that digital printing has great applications in the production of packages. Many desktop digital printers now have envelope printing capabilities.
- **Financial/legal.** This category is composed of quarterly and annual reports, 10Ks, and investment/legal reports. Not unlike the brochure/booklet category, these can be produced on-demand.
- **Invitations/menus.** Invitations are relatively simple, but menus can be quite complex, but they both require very short runs.
- **Letterhead/stationery.** Letterheads and stationery consist of flat, single-sided pages. This category also includes business cards, tags, and labels, which we expect to include more color. You may view the machines in airports and malls, where they make business cards and stationery as on-demand systems.
- **Newsletters.** This is one of the most popular applications of desktop technology (books/manuals being first). Most newsletters have smaller-than-average run lengths, and are prime candidates. This has become a popular application for the Xerox DocuTech technology.

- **Signs/posters.** These products are categorized as display-oriented material and are characterized by large type sizes and increasing use of color. Digital printing is usually applicable up to 11×17 in., which covers a significant amount of single-sheet work.
- **Books/journals.** Journals may be more apt to go on the Internet or CD-ROM, but we envision a variety of approaches to on-demand book production. See the chapter on this subject.

The list that follows has over fifty products that lend themselves to on-demand printing. It continues to grow as early practitioners use the imagination to meet customer needs. The chart on page 49 shows the major categories of short-run printed products by primary application.

| | | | |
|---|---|---|---|
| Data sheet | Stationery | Pre-print | Menu |
| Fact sheet | Meeting notes | Reprint | Operating manual |
| Product specification | Book cover | Presentation | Greeting cards |
| Promotion sheet | Manual cover | Notice | Guide |
| Price list | Record, CD, video cover | Newsletter | Organization chart |
| Counter card | Packaging | Small journal | Directions |
| Folder | Instruction sheet | Poster | Map |
| Conference program | Flyer | Signage | Parts list |
| Trade show handout | Brochure | Art reproduction | Custom book |
| Package insert | Personalized check | Coupon | Real estate promotions |
| Label | Sales letters | Retail material | Catalog sheet |
| Form | Direct mail | Wholesale material | Custom catalog |
| Invitation | Ad | Report | |

## Inappropriate Applications

If you look at others people's lists of what you can do with on-demand, digital, or customized printing you almost wonder — what can't be done? Perhaps we should list those products that do not lend themselves to short-run approaches. The question was simple: what kinds of products would continue to require long runs? Here is our list:

1. **Consumer product packaging.** Once past the design and testing phase, most mass-market products would require runs in the millions. Theoretically, you could make a case for regionalization of production or even smaller runs, but we just can't see packaging a few hundred boxes of Corn Flakes™, unless they were making a special run for you with kumquats and raisins.
2. **Metropolitan daily newspapers.** Whatever you say about the newspaper industry, the metro daily will still be a mainstay, if only to give commuters something to do. There will probably be fewer newspapers, but there will be newspapers.
3. **Mass-market books.** Certain authors and their books will be able to sell millions of copies in either hardcover or softcover form, and it makes sense to print in longer runs.
4. **Political and institutional fund-raising promotions.** Blanket mailings to every home or selected homes are the only assured way of reaching a mass audience.

## Major Categories of Short-Run Printed Products

| Periodicals | Books | Catalogs | Technical Documents | Direct Mail |
| --- | --- | --- | --- | --- |
| Journals | Reference | Business | Manuals | Letters |
| Newsletters | Textbooks | Industrial | Guides | Reply cards |
| Reprints | Workbooks | Dealer | Parts lists | Post cards |
| Preprints | Technical | Distributor | Maintenance | Notices |
| Comics | Juvenile | Retail | Repair | Coupons |
| Magazines | Yearbooks | Specialty | Application | Self mailers |
| Newspapers | Out of print | Salesman | Ltd. production | Inserts |
| Shoppers | Graphic novels | Targeted | Instructions | Variable offers |
| Customized | Customized | Customized | Customized | Customized |

| Directories | Promotion | Legal (Advertising) | Packaging Financial | Other |
| --- | --- | --- | --- | --- |
| Membership | Flyers | Quarterlies | Labels | Reports |
| Governmental | Brochures | Notices | Book covers | Proposals |
| In-house | Folders | 10-Ks | CD covers | Presentations |
| Schedules | Booklets | Checks | Record covers | Certificates |
| Attendee lists | Data sheets | Prospectuses | Test packages | Menus |
| Professional | Posters | Investor info | Tags | Greeting card |
| Telephone | Countercards | Legal statutes | Displays | Policies |
| Price lists | Ads | Annual reports | Signage | Stationery |
| Customized | Customized | Customized | Customized | Custom form |

*Source: NEPP, Digital Printing Report, Vol. 1 No. 7, 1994.*

5. **Tax forms.** Although the government will encourage electronic filing, the majority of taxpayers will not have access to the technology.
6. **Telephone books.** Eventually, the cable system will link your telephone and television, and the directory will be electronic.
7. **Certain magazines.** Many general and special interest magazines will still have vast numbers of readers for the print version, even if they offer an electronic version.
8. **Certain catalogs.** There will be the need for mass distribution of certain consumer catalogs, especially if the products appeal to a large cross-section of the population.
9. **High-volume direct mail.** We could make a case that even though it might not be delivered by the Postal Service, direct mail will exist for certain mass mailings, especially the ones that tell us that we could win millions of dollars.

10. **Promotional material.** Brochures, flyers and the like will be used by dealers and distributors that have products and services for a mass market.

There are probably others, but these came to mind quickly. Let us not assume that all reproduction will go short-run. It will certainly be digital, but run length depends on the audience.

# Chapter 7

# Book Publishing

Publishing books, like other niches in printing and publishing, is changing based on emerging technologies and ancillary production issues such as paper cost, inventory risk, and distribution costs. As a result, some publishing companies are merging or developing alliances with communication and entertainment companies. For example, Viacom Inc. owns Simon & Schuster publishing unit as well as Paramount Communications Inc. and Blockbuster Entertainment Corp.

Developing more efficient production methods for creating publications and more efficient delivery vehicles is becoming more and more important as publishers compete for consumers' attention with other media.

Due to the alternative production and distribution options, the methods used to prepare, create, and distribute traditional books created on an offset printing press using film-based prepress, stored in warehouses, and distributed with ground-based methods are currently under fire.

Traditionally, book publishing has been accomplished using offset printing with the run lengths based on estimated demand or cost efficiencies. Due to the costs associated with the printing process, especially the prepress and press makeready costs, publishers typically produce books in run lengths of 1,000–10,000 units. Once printed, the majority of the books are warehoused and distributed as demand dictates.

Alternative books would include CD-ROM, online, and books created on on-demand presses or digital color presses. With information content, teaching styles, and classroom demographics changing annually, can that stale old black-and-white textbook maintain its value as a teaching tool? Furthermore, how will it compete with color books or books on CD-ROM with interactivity and audio/visual capabilities?

## Issues in Traditional Book Production

The traditional methods of creating books are costly because of the prepress costs, which include film, processing chemistry, and plates, as well as the time and cost of collating and binding. In addition, high costs due to setup and spoilage in both the press and bindery areas increase the costs on short-run books and first editions. This problem is most apparent in specialized academic and professional fields, where a popular title may only sell a few thousand copies.

Another expensive operation in book production is collation which is the process of gathering the pages into the correct order. In book publishing it is often a greater expense to collate after the printing rather than before. This can add additional expense to the final price.

One factor particularly relevant to book publishers is the risk of obsolescence. This increases the advantages of on-demand printing. Textbooks used in high schools or colleges can become obsolete in only one or two years. For this market segment, producing textbooks on-demand would reduce costs and risks.

Another factor that has placed traditional books at a competitive disadvantage was a 1980 IRS ruling called the "Thor Power Tool" case. Before this ruling, publishers used to print several years worth of slow-moving books and store them in warehouses until they were needed. Using this strategy they could print enough books to achieve a cost-effective pressrun. However the "Thor Power Tool" case forced the publishers to account for inventories at their full list price for tax purposes. The result was a strong incentive to decrease inventories by decreasing print runs, which increased the per-unit cost.

For most publishers the bottom line is more important then ever. In an interview in *The Horn Book* magazine (Jan.-Feb. 1995), William Morris, who is involved with library promotion and advertising at Harper Collins, said "The bottom line is more important today. Many houses are not as willing to take gambles... Now many seem to prefer getting someone who's already established. They want immediate hits."

Marketing factors also influence the viability of books created with traditional procedures. For example, the university and school book markets are affected by the used book market. Students often prefer used books to new books because they are cheaper and the text is already highlighted.

Obsolescence is an important word in book publishing. Obsolescence means that the product is no longer useful for the purpose for which it was created. Slowly, and over an extended period of time, what once was a timely and appropriate item becomes out-of-date and worthless.

There are different rates of obsolescence. Some books like *Tom Sawyer, Catcher in the Rye,* and *Treasure Island* may never lose their value. The content in these classics may never be altered. On the other hand, a large majority of books do suffer from obsolescence.

For these books the ability to update becomes very important. Any books that discuss technology require frequent updating.

## Customized Textbooks

For the 3,300 colleges and universities in this country, the costs of printing short-run documents has skyrocketed. Besides the course materials, these documents also include manuals, research reports, and alumni materials.

In addition, with the average publishing and production cycles ranging from 18 to 36 months, the textbook industry is hard-pressed to keep pace with the demand for rapidly changing information. This is most evident with areas such as computer-related or technology-driven research.

As a result, teachers are challenged to find current sources for classroom subjects and sometimes create their own course materials. These are customized materials compiled from chapters of separate textbooks, articles, essays, and other materials. Often, the issue in creating customized books is permission for the use of copyrighted material. As we will see in subsequent sections, these can be handled through college bookstores or off-campus copy shops.

## Issues and Answers with Customized Textbooks

With the price of paper and printing rising, universities and other schools are challenged by several problems with course material. How do these institutions make their materials more timely, accessible, and affordable. In addition, schools are looking at ways of managing and customizing instructional materials. These institutions have begun to realize the potential for digital document management.

For years, college professors have created customized course notes for their students. Called Professor Publishing by Kinko's, this unique product, service, and market niche caused the explosive growth of Kinko's shops nationwide until they were sued for copyright infringement.

In Professor Publishing, the professor chooses chapters from differently

At GATF, we use customized books to create courseworks for our workshops.

books as well as recent journal articles and creates a customized textbook. Then either the print shop or the university in-house reprographics department or the publisher prints just enough for the enrollment in that professor's course.

There are strong motivations for both students and faculty. The advantage for the students is that they don't have to buy several books and use each one on a limited basis. For professors, they can take information from many sources and update the material quickly and easily.

The market first used by Kinko's copy shops remains viable with only one caveat — the copyright issue. Kinko's was sued because they did not pay for copyrights and subsequently de-emphasized this service. But the copyright issue can be successfully overcome. For example, Kevin Schostberger from the University of Chicago Printing Office can print over 150 different course packs every night during enrollment and is a stickler for copyright permission.

## Copyright Issues

Kinko's Graphics Corp. lost a two-year battle in U.S. District Court in New York on March 28, 1991 when the judge found that Kinko's infringed on publishers' copyrights in selling photocopied course packets to students as study aids. The lawsuit was initiated by the Association of American Publishers (AAP) and eight publishing companies. The court ruled that Kinko's was infringing on copyright protections when its copying shops around the country sold excerpts from books to students without obtaining permission or paying royalties.

In an interview in the *Columbus Business First Newspaper* (Sept. 16, 1991) Ken Zaeger, coordinator for permissions for MacMillan Publishing Co. College Division, said the ruling will change the way copy shops do business. "People are going to be more careful," he said. The number of requests his office is processing has changed dramatically. "Last year we did 500," he said. "Now we're doing 500 per month. I'm so backlogged, it's unbelievable."

Speaking at the Textbook Authors Association shortly after the court decision, Kurt Koenig, vice-president of Kinko's Service Corp., said that Kinko's was replacing Professor Publishing with a new service called Course Works.

Koenig predicted that customized publishing would change the way royalties are paid at all levels, and called the current system for obtaining permissions and paying royalties "messy." He also stated that administrative costs for publishers, as well as for the Kinko's organization, exceeded the amount of royalties paid.

He suggested that an alternative system should be devised in conjunction with the Copyright Clearance Center and the AAP. He also suggested that a licensing arrangement similar to that used by ASCAP (the American Society of Composers, Artists, and Producers) might be a way to eliminate overhead costs.

## The Copyright Clearance Center

One possible solution is the Copyright Clearance Center in Massachusetts, which is becoming very active in administrating copyright issues for course pack materials at college bookstores around the country. Requests are growing in leaps and bounds. In August 1993 it handled 40,000 requests for copyright clearance of materials—an increase of 148% compared to 1992. In 1994 it distributed more than $2 million in royalties.

Hopefully, some new systems will help with the administration. According to CCC's director of professional relations, Isabella Hinds, CCC recently agreed to an alliance with Cornell University whereby the CCC database will be accessed through the new Xerox OMS (Operations Management Systems) software. This streamlines the system and automatically records all copies made. Cornell can then send those records and the money, collectively, to CCC, which will distribute it to the appropriate publishers.

"We will have to make an agreement like this with each buyer of the Xerox OMS software," Hinds admits. "But it will streamline the process enormously, and our purpose is to help keep costs down and compliance up."

CCC has restricted this service to Xerox. It is working with other electronic course pack services as well as with proprietary systems such as CAPCO (a joint venture of Follett and BMI) and Barnes & Noble, and with larger campus bookstores, such as Cornell and Stanford, which are developing their own systems.

In addition, CCC has signed an agreement with a number of textbook publishers in the Paramount group, which will streamline copyright clearance of materials at a flat per-page rate. With this kind of cooperation from publishers, the success of electronic delivery and local production of materials around the country and around the world is assured.

## Book Distribution: Issues and Answers

There are a number of issues involved with book distribution. For example, in May 1994, the American Booksellers Association (ARA), which represents 4,500 independent book sellers, filed suit in U.S. District Court in Philadelphia against five publishers, charging that they violated antitrust laws by giving preferential price breaks to chains and warehouse clubs. According to the ARA, unit sales in independent bookstores dropped to 25% of all books sold in 1993.

A more relevant issue to the subject of this book is that the cost of distribution is high when compared to the costs of creating and distributing books with alternative methods such as an on-demand, CD-ROM, or online book.

Traditional books are costly, but much of the cost is not due to production. According to Tony Rothman a Harvard University professor, for a book costing approximately $2— 40%, or $8, will go to the retailer. Of the remaining $12, about $3 will go to the distributors who ship the books from the printers to the bookstores. Of the $9 balance, it might cost $3 to produce the book. The author receives $2, and the publisher keeps $4. All of these numbers vary considerably, but as a rule of thumb, a book's cover price is six to eight times the production cost.

According to Mary Lee Schneider, former director of marketing for the R.R. Donnelley Digital Division, between 30% and 50% of the cost of a book is associated with inventory maintenance, inventory risk, and distribution. Using traditional strategies today, distribution and inventory management can account for as much as 30% of the overall cost of producing a printed product. On-demand printing can eliminate most of those costs. Customers could save almost 30% by avoiding inventory maintenance, distribution issues, and associated costs for inventory risk.

## Kodak/McGraw-Hill Primis Project

One of the potential advantages of alternative books is distribution methods. One way to decrease distribution and warehousing costs is to save data electronically in a central location and telecommunicate the pages to regionally located on-demand presses. The first attempt at this strategy was the Primis project, developed jointly by McGraw-Hill's (publisher) specification by Kodak (on-demand press manufacturer) and R. R. Donnelley (world's largest printer).

The Primis project was the first system to offer custom textbook production. The publisher licensed Primis software to college bookstores and campus printing facilities, enabling them to establish local publishing centers.

Primis products are both cheaper and more elegant than those generated to date on the competitive product, the DocuTech, especially when created at an R. R. Donnelley site, where the majority of project orders are sent. Primis charges 4–6¢ per page, including copyright and discounting for products longer than 400 pages.

In 1991, students and faculty at the University of California, San Diego campus (UCSD) were among the first to use the Primis project. Schools using the Primis strategy allowed professors to look through a McGraw-Hill catalog and order any journal excerpts, articles, and reviews pertinent to their course. After the professor decided on the articles, they were given a bound copy that included the articles, a table of contents, an index, and a unique ISDN number. If the professor approved the publication, the book was produced in precise quantity and delivered directly to the college bookstore for student purchase.

Primis offers advantages to campus book stores as well. On-demand publishing and digital color printing of textbooks allows instructors to order only what is needed for each class. Additional copies, even one or two books, can be supplied within 24–48 hours. Bookstores will avoid the cost of carrying overstocked inventory, and shipping the excess back to publishers for full or partial refunds.

Two stumbling blocks have affected the success of Primis, however. One is that professors want the ability to order from any publisher. More importantly, however, is that most major textbook publishers have refused to put their materials into the Primis database. The main reason given is concern over rights — both intellectual and copyrights.

The publishers argue that the integrity of their materials will be compromised if they are "cannibalized" into bits and pieces. Typically a professor will want only certain chapters of a particular textbook and will combine that with other materials.

## Primis from a Publisher's Perspective

Addison-Wesley is licensing the Primis system while continuing to use its own proprietary electronic database. The company works directly with the Primis print partner, R. R. Donnelley, at its plant in Pennsylvania, according to Ann DeLacey, vice president of corporate manufacturing/inventory management East at Addison-Wesley.

In an article in *Publishers Weekly* (June 13, 1994), DeLacey is quoted, "We've integrated our own royalty tracking and so forth, and enter orders directly from our customer service/order entry group to the order entry system at the prepress facility that Donnelley uses for fulfilling the Primis orders. Our regional sales managers had expressed reservations concerning problems in product quality and delivery that were reported by end users of the early Primis product. We worked with the R. R. Donnelley management to ensure that Primis could meet the needs of our customers."

In projecting the future of demand publishing DeLacey said, "I can't yet say what the full impact of custom publishing will be for A-W. Our product lines are mainly hard science and math texts, and therefore linear in content and pedagogy. It can be difficult to break and rearrange chapters or sections without harming the pedagogical content of the product. Each project is evaluated on an individual basis. The current system and equipment limit us to one-color or to spot, cosmetic two-color. However, the new four-color reprographic machines currently being tested may have large applications for customized products or limited-run situations, i.e., school adoptions, where you generate a small first-print quantity and do major corrections before film and traditional longer print runs."

## Xerox Courseware On-Demand

Xerox has also seen the potential application for professor publishing. Xerox has a solution that is called *Courseware On-Demand*. It is an integrated hardware and software system designed especially for academic publishing.

The components of Courseware are delivered by Xerox Documents On-Demand (XDOD) and Operations Management System (OMS) software. It has the ability to scan course pack originals at a PC using the Xerox DocuCM 620 scanner and store the digitized originals on a jukebox server. You can index, search, and finally print your documents on a DocuTech Publisher or DocuPrint printer.

The ease and speed of printing extra sets reduces the need for costly overruns. The course pack files can be stored electronically, recalled, updated, and reprinted on-demand. The system's copyright management software automatically generates permission requests, tracks responses, and calculates royalties.

Course packs can be finished in-line with tape binding, perfect bound with the new Bourg Binder BB2005, or off-line with a variety of Xerox and third-party finishing products.

Course pack covers can be printed on a 5775 SSE Color Copier/Printer with Fiery 200i Controller. They can also be produced with a variety of Xerox high-speed printers; one configuration integrates the DocuTech Network Publisher and Network Server. The Network Server runs TCP/IP software and Novell protocols link the document creation and storage system to the production printer, enabling distributed printing at remote locations. Documents can be stored on the XDOD Server with HP optical jukebox. Documents are scanned at the XDOD mastering station with a DocuCM 620 scanner.

## Summary

In summary, on-demand, digital, and customized book publishing offers very powerful advantages over traditional book publishing. These advantages include:
- Short-run production: print only when you need it
- Reduction in inventory, inventory risk, and storage costs
- Elimination of obsolescence, because books are easily updated

These solutions can be utilized by various groups in our industry. Currently these groups include:
- Publishers who want to print books on-demand
- Campus bookstores or off-campus copy shops that create customized course ware
- Libraries to preserve and store books digitally instead of storing them on shelves or on microfilm

The benefits to customers include:
- Improved economies for short runs
- Fast response to customer orders
- Books will never become out of print because you can always print an original from the digital file
- Books would never have to be deleted from a catalog unless the content becomes obsolete
- Authors' proofs and review copies could be produced in record time
- Color pages can be "tipped in," and color covers printed on color copiers or other short-run methods
- Short-run books can be produced cost-effectively
- New markets reached through publishing titles not possible due to the economics

# Chapter 8

# In-House Applications

One of the advantages of desktop computers and publishing is that the low entry-level costs made document creation and publishing more accessible for the masses. The masses ranged from students using Apple II computers to professionals working in-house centers preparing documents for printing.

Large companies depend on documents for everything. There are often large support systems supporting in-house authoring, printing, and distributing of paper documents. In many cases, they create the fabric of communication within a company. Besides their role in internal communication, many companies depend on document creation for advertising and documenting the completion of individual tasks (i.e., sales).

Traditionally, there have been two reasons to create an in-plant printing services — cost and control. Recently, however, a new motivation has emerged, the motivation for inexpensive color documents in short runs.

## Color Documents Are HOT

The use of internal color documents is increasing. A recent study from the Hewlett-Packard Company suggests that a majority of American corporations is moving to color printing. In the survey, 75% of the 400 MIS managers queried said that the businesses they support have acquired or plan to acquire color printing capabilities in the near future.

The survey, which was entitled "Color Printing — An Emerging Need for Persuasive Business Communications," found that almost two-thirds (63.7%) of the businesses that have color printing capabilities expect their use of color printing to increase in the next two years. Of those planning to enhance their color capabilities, more than half (58%) said demand for color printing within their companies is increasing and that they need a greater number of color printers in the office.

According to Norwell, Mass.-based market research firm, BIS Strategic Decisions, an overwhelming 96% of respondents with color printers agree that color printing makes business communications more persuasive, compelling, and memorable.

Eye-catching, internally-produced color is being used to call attention to everything from computer-generated designs to presentations and internal and external documents. If companies with internal color devices decide to increase the run length from a handful to a few dozen, then they will realize they need devices faster than the 4-page-per-minute color copier.

## Cutting Costs

In business today, everyone is looking for ways to cut costs and increase profitability. Almost everyday an article appears in the newspaper describing how companies are eliminating nonessential business services and costs in an attempt to increase cost-effectiveness.

Progressive organizations are building competitive advantages by delivering documents faster and cheaper. This ability increases sales effectiveness in indirect ways as well as direct ways, analogous to how reengineering increases productivity in direct and indirect ways.

Companies that utilize digital and on-demand printing are increasing the speed of communication. This improves the businesses' responsiveness, increases productivity, and lowers total operating costs.

The advantage of on-demand and digital color printing that makes it so attractive for in-house operations is that the information remains digital until the moment before it is needed on paper. This allows data to remain fluid and insures that the most current information is printed.

U.S. businesses spend over $180 billion a year on printing. How much a company spends depends on the specific company, the product or service it sells, and the company's size. Studies show that the costs are much higher than companies estimate.

There are two factors commonly associated. One is the failure to recognize the total cost of production. Often, the corporate focus is on cost per page and not the total cost for delivering the printed piece. Factors often neglected include storage, retrieval, document tracing, shipping, disposal, and obsolescence.

The second factor is an inaccurate, underestimation of the amount of printing done by the in-house operation. According to companies that specialize in in-house organizations such as Xerox and Interleaf, 90% of the printing in a company is done outside of the plant.

## Controlling Mission-Critical Data

One of the most important uses of documents is for the communication of mission-critical information. Another way to discuss them is to categorize them as strategic. They may include new drug applications in the pharmaceutical market, product reference manuals in the high-tech industry, or product portfolios in an investment company. In contrast to office memos and database information such as bank statements, they are strategic because they contain the company's most critical information.

Documents that contain business-critical data are at the very heart of these business procedures. To take advantage of this asset, organizations are reengineering print production with document management systems that allow them to store, access, distribute, and manage documents efficiently.

Using traditional print production methods requires a long period of preparation and a hand-off from the document creator to the print shop for scheduling. After the hand-off, the print shop requires lead time for setup and print production. Printing business-critical documents on-demand decreases turnaround time, reduces cost, and increases timeliness.

Two interesting examples of mission-critical information follow. Two consulting companies, Ernst & Young and Andersen Consulting, depend on the communication of mission-critical information internally to compile reports, and externally for timely delivery of information.

## Case History: Ernst & Young

Ernst & Young operates in more than one hundred countries worldwide, and is the world's largest professional services firm specializing in accounting and consulting services. Ernst & Young's market focus is on financial services, health care, high technology, manufacturing, and retail industries.

As a management consulting firm, Ernst & Young has expertise in business process reengineering and image management. The firm maintains its worldwide presence through a network of more than 5,000 consulting and support services personnel.

Xerox's initial alliance efforts with Ernst & Young focused on document-based process reengineering for the insurance and financial services industries. The alliance focuses on developing unique and creative strategies to improve a client's organizational effectiveness and maximize customer value through application of innovative document processing solutions encompassing people, process, and technology.

The basic framework of this concept is the strategic significance of documents within key business processes. It aims to develop business process solutions that are applied within document processing environments.

Ernst & Young has installed five networked DocuTechs at its National Supply facility in Cleveland and supports Xerox Document Production Systems technologies as a key element of the partnership.

## Case History: Andersen Consulting

Andersen Consulting is a business unit of the Andersen Worldwide Organization and the world's largest information systems consulting firm with 25,000 employees and 229 offices worldwide. Andersen Consulting provides professional services in all aspects of information processing in virtually every industry with special focus on government, healthcare, manufacturing, banking, and insurance. Andersen enjoys high client privilege, support, and loyalty.

Andersen Consulting is a worldwide leader in systems integration and change management. Xerox products and services are leaders in process reengineering. Together the combination is perfect for the client who demands consultative expertise and end-to-end implementation of improved document management techniques.

To demonstrate these capabilities, Andersen has implemented a Just-In-Time Print Center in Chicago. This center is an excellent example of how a traditional offset print-and-store facility can be converted to a print-on-demand digital environment. Extensive savings have been realized with reduced delivery time to users. This location is a model for executive commitment, study technique, and cost justification for in-plant printing in the 1990s.

## Case History: Healthcare Forms

In the several areas within the healthcare industry, the need to decrease costs and improve service has become critical. This is true for hospitals as well as managed healthcare organizations. One of the ways to accomplish this goal is by using on-demand and digital color printing.

If you've paid any hospital or doctor bills lately, you realize that the U.S. health-care industry is undergoing radical changes due to the ever-increasing cost of medical insurance and hospital costs. Given all of the pressures on the healthcare industry, the main objectives of most healthcare organizations today are to increase the quality of healthcare services and to extend and expand access to services while reducing operational costs.

Administrative costs in hospitals can reach 25% of operating costs due a large extent to the cost of printing directories, manuals, and reports that are only needed in small quantities and only have a short shelf life.

These hospital forms are a perfect application for on-demand printing. By improving the way in which forms, directories, manuals, public health booklets, legal briefs and contracts, presentations and reports, brochures, newsletters, flyers, and menus are printed, stored, and distributed, hospitals can help make a substantial difference in healthcare costs and delivery.

Managed healthcare organizations all over the country, from HMOs to Blue Cross/Blue Shield organizations to hospitals, are struggling to give their members provider directories that are accurate and up to date. The problem is that current production processes require lengthy preparation time, endless correction cycles, and large print runs that result in higher costs.

Faced with skyrocketing expenses, eroding revenue rates, and impending government reform, U.S. healthcare organizations share a mandate to cut costs. But every year, they are forced to throw away an estimated $2 billion on forms that become obsolete before they are ever used.

With traditional copiers or offset printing, forms production is slow, costly, and complicated. Traditional copying methods using paper masters do not allow for easy revisions — a big disadvantage, since forms must be updated frequently due to changing government and insurance regulations. Long prepress setup times necessitate large offset print runs, forcing healthcare organizations to print more than they need, warehouse what they don't use, and throw away inventory when forms need revision.

Using on-demand and digital printing services, HMOs are creating an integrated system designed especially for the managed healthcare industry. It reduces the length of time necessary for gathering and organizing information for printing. It shortens the update cycle time by allowing electronic editing, compilation, printing, and production.

Plus, you can print the quantity you want, when you want, thus cutting the cost of production, mailing, inventory, and waste. The flexibility of this system helps reduce costly overruns by allowing you to print on-demand. Directories can also be customized to serve a targeted group of subscribers.

## Case History: Northern Trust Bank

The in-house printing department at Northern Trust Bank is fighting a battle. The goal is to keep work in house, and recover work that has been sent to outside suppliers in recent years. In many ways, it's the same struggle faced by most printing departments at large organizations. At Chicago-based Northern Trust Bank, the in-house department is winning. It's winning not because it offers reduced costs, although in many cases it does, but because it offers superior service in the form of greater convenience and rapid turnaround.

The key to the department's success is using technology wisely. Ruth Johnson, vice president for publication services, is responsible for merging the bank's computer systems with new electronic printing technology. Northern Trust embraced the idea of high-volume computer printing services back in the late 1980s. Today, it operates two Kodak 1392 printers, each capable of printing 92 pages per minute. They are driven by any combination of computer platforms — from Macintosh and IBM-compatible desktop computers, to a Sun SPARCstation 10 with Interleaf software.

Tying the enterprise together is Kodak Lionheart document imaging software. This allows the printers to accept input from virtually any computer platform and create documents produced in any page description language, which positions Northern Trust to supply on-demand printing. Documents can be stored electronically by individual users in departments served by the printing department, or on a file server in the publications department. In either case, documents can be updated at any time, printed as needed, and distributed throughout the bank or to bank customers. That eliminates waste because the bank does not print hundreds or thousands of extra copies of a document, only to throw them away when they become outdated. It also ensures that documents are current when they are printed, and it reduces production time.

## Case History: Hughes Aircraft Co.

In 1989 Hughes Aircraft Co. of Long Beach, California, implemented an on-demand press as part of a reorganization effort aimed at reducing costs and increasing productivity. The company's Art Services Department purchased a desktop graphics-creation system, which enhanced its ability to develop eye-catching and sophisticated publications in-house, and the Copy Department bought a DocuTech Production Publisher.

At the time Hughes had 67,000 employees in 21 countries. The goal was to overhaul the management of its tremendous paper production without adding staff. Despite a near-100% increase in copying volume over the next two years (from slightly over 10 million copies in 1989, to a projected 20 million in 1991), Hughes didn't incur additional expenses.

The only changes were in equipment and supplies, as staff and operating space were not increased. Hughes expects to continue in its efforts, anticipating DocuTech additions that will enable the Production Publisher to be networked with desktop computers.

## Case History: Insurance Companies

Insurance company printed products include policies, benefit booklets, rate manuals, and other informational media. Because insurance companies are heavily dependent on documents to interact with customers and to communicate internally, and since government regulations dictate that paper documents serve as the legal conveyance of information, paper flow within most insurance companies is a major contributor to costs and to the ability to service agents and clients.

In an effort to reduce costs and improve turnaround time, many insurance companies have invested in desktop publishing systems. To realize the promised benefits of electronic storage and retrieval of documents, and to improve overall competitive advantage, insurance industry leaders are moving to on-demand and distributed printing for a host of critical business applications.

Using on-demand and digital printing equipment allows insurers to cost-effectively produce customized health benefits books for small or large groups. Representatives can create these books at their desktop and order any number of sets.

One of the problems in the insurance business is that the demand for forms is unpredictable, which results in long runs, storage, and possible obsolescence. Creating and storing forms electronically allows for faster and easier updates and maintaining a lower inventory.

The advantages for this application include more timely delivery, quality documents satisfy existing clients and attract new ones, lower costs, and the ability to customize or personalize cover letters and benefits books allows the company to stand out from the competition.

## Marketing

In terms of attracting new customers, insurers have an urgent need to promote the healthcare providers included in their plan. In the past, this was accomplished by having the company distribute large, expensive directories. The problem is that recipients, especially some of the older ones, found it difficult to search through long lists of names to find physicians in their area.

The answer was combining database functionality with on-demand digital presses. This way the insurers could organize their lists of providers by zip code and send out smaller, less-expensive, easy-to-use booklets that contained providers located in the recipient's immediate area. Using more elaborate database procedures, some insurers customized the documents by including black-and-white photos of physicians, along with a brief description of their specialties.

One major insurer already encourages its agents to create reports, direct mail, or other customer-oriented materials at a desktop computer and print them from centralized regional offices, transmitting files, complete with desired distribution lists, to the in-house printing system. Name and address data is merged with the documents to create personalized communication, and jobs are automatically output and mailed.

Using electronic job tickets permits individuals to specify finishing and delivery instructions directly into the job ticket without making a phone call or leaving their desks. As a result, productivity is greatly improved throughout the entire organization, not just in the central office. Using on-demand technologies allows insurers to adapt quickly and easily to changes, such as fluctuations in procedures, regulations, and premium rates. Traditionally, this would require additional print runs to constantly update training and procedure manuals as well as agent rate books.

For insurers, traditional methods of print production are time-consuming and expensive. For them, on-demand technologies are more flexible. They can produce and distribute updated materials in days, not weeks, and forms can be revised and updated quickly and inexpensively.

## Case History: Manufacturers

Manufacturers are facing several challenges today. Foreign competition is having a devastating impact. In addition, manufacturers are facing government regulation, changes in consumer buying trends, and the rapid pace of technological change. As these challenges work to reduce the life cycle of many products, and increased the importance of just-in-time manufacturing, they have also altered the way in which manufacturers produce products, conduct product development, and create support documentation.

Manufacturing companies recognize the power of information. It is not enough to make products quickly; companies must make the right product, release it at the right time, and market it to the right audience. Information must be shared by engineering, manufacturing, sales, marketing, and management.

Since market dynamics change each day, timeliness is essential to accurate decision-making. These companies require a system that quickly collects, compiles, and serves information to users — and provides equally speedy delivery.

With the current state of desktop systems, manufacturers can improve production, storage, and distribution of mission-critical documents — from user manuals, maintenance guides, training materials, and parts lists to work orders, illustrated assembly instructions, engineering specifications, and directories.

## Retail and Wholesale Applications

In the retail and wholesale markets, competition is at an all-time high. There is pressure to increase market share, revenue, and profits, to track buyer trends, to improve customer service, to establish strategic alliances, to build or strengthen value-added services, and to reduce costs. To answer these concerns, management is looking for well-defined solutions to their document management problems.

On-demand can help retail and wholesale companies improve productivity — not only in document creation but in facilitating decision making — by getting the right documents to the decision maker at the exact point of need. Well-integrated on-demand systems can eliminate steps in the offset production process and reduce costs and improve turnaround time, enabling them to respond to niche demands and better meet the requirements of customers and suppliers.

Downsizing or rightsizing has caused these companies to examine more economical printing technology. These companies, like others, have found that improving efficiencies in storing, distributing, and printing documents can increase productivity especially in the areas of forms automation and publications production.

## Summary

There are several good examples of in-house on-demand solutions that have been successfully applied such as financial and professional services, government, healthcare, forms management, and wholesale/retail.

Documents have an essential role in businesses that serve the financial and professional community, including insurance companies and banks. The documents created by these organizations are the vehicles for the services they provide to their clients.

Growth, combined with the rate of information change and the desire to keep consumers informed, has made it very difficult for organizations to create, store, and distribute up-to-date documents. The documents required are close to 100 billion pages a year. Of those, 10 billion are handled in in-plant agencies, and 60 million pages performed through outsourcing. As control and speed of delivery becomes more important, more corporations will bring these services in-house.

Forms printing is the largest printing application in the healthcare industry. For example, hospitals with 300 or more beds use 1,500–5,000 different forms. The resulting printing and publishing volumes per hospital can range from 280,000 to 900,000 pages per month. This is expensive and slow. Forms are either bought or produced in-house on offset printing equipment. Usually long printing runs are performed, and the forms are warehoused.

Since forms are constantly updated due to changes in governmental regulations, insurance policy changes, hospital requirements, and record keeping, the old forms become obsolete and are thrown away. Costs are increased as revised forms are printed and stored to replace the obsolete forms.

Although not discussed earlier, another application for this technology is for national, state, and local governmental agencies. These agencies have diverse documents requirements. As a result, often it is difficult to anticipate the number of documents required. For example, some issues generate great public and press interest and a concomitant number of documents while others do not. As a result, it is not uncommon to print too many documents, which are thrown out, or too little, which require additional printing.

# Chapter 9

# Label Printing Markets

Customer demand within the label printing industry, like book publishing, forms printing, and other niches in print production, are changing. Central to the new customer demands in the label printing industry are shorter run lengths, increasing customer demand for higher quality, and more four-color process printing.

While customer demand is shifting, other changes make on-demand and digital color printing more attractive than traditional printing solutions. One of these changes is a growth across the board in the label marketplace. Other factors include the increased use of multiple labels on single packages as well as an increasing popularity of data merging and variable data. This chapter examines the label markets.

As a result of the wide variety of subjects included in this chapter, a large number of acronyms have "grown" into this chapter. In addition, the label markets and applications can be somewhat confusing. At the end of the chapter, we have included a list of acronyms and label types.

## Introduction to Label Printing

Most equipment and production strategies in printing and publishing are similar in terms of the technology and utilization, but the label industry has unique problems and concerns. Most of the printed label products used today are for packaging. They are affixed temporarily or permanently to products, to identify contents, provide use instructions or cautionary information, and/or improve the sales appeal of the item.

Labels are universal in the world of packaging today. Labels range from those used to promote and sell the product ("prime labels"), to those used to provide instructional or cautionary information to the user or consumer, to those used to protect the product or to guarantee product integrity (these labels are called "secondary labels").

Not all labels are used in packaging. Labels are also used in various applications in data processing, in offices, and at home. Labels are used to address envelopes and produce return addresses, to reward elementary students for good performance, to identify people at social functions, to price products at retail stores everywhere, to identify products with bar codes and alphanumeric symbols in industrial applications, and for a myriad of other non-packaging purposes.

As a general rule, labels used in packaging are more complex, more expensive, and more intricately designed than those used outside of packaging. Labels — especially the ubiquitous self-adhesive label — permit putting information directly on a product or device without permanently changing the character of the item. Various adhesives may be used, ranging from those that permit easy removal to those that cannot be removed without destroying the label. Specific adhesives are available for use in harsh environments and for use on products that may be easily damaged.

## Label Types

The principal label product areas include the following:
- Pressure-sensitive labels
- Labels with moisture- or chemical-activated glues
- Sheet labels, without adhesive
- Heat-shrunk plastic labels
- In-mold labeling

Pressure-sensitive (PS) labels range from the information and cautionary labels on an individual's prescription bottle to grocery store item labels, and from personal mailing labels to a myriad of industrial uses. Pressure-sensitive labels may be printed in advance and applied to products as part of the production process, or may be generated one at a time to convey information unique to a particular package.

The widespread use of laser printers and personal computers has increased the use of pressure-sensitive labels in low-volume packaging applications. Pressure-sensitive labels may be found on products of virtually every type and kind.

Labels with conventional adhesives are those activated with heat, moisture, or chemicals and are still widely used, despite the rapid gains in market share being enjoyed by pressure-sensitive labels. Applications include industrial labeling of various sorts, as well as consumer packaging.

Sheet labels are those used on canned foods and are typically attached to the can or bottle. The substrate used may be paper or plastic, and it may cover a wide range of weights and grades.

Heat-shrunk plastic labels are relative newcomers to the label segment. These labels are found on soft drink and similar bottles.

In-mold labels are permanently bonded to a container during its manufacturing process. These labels are available for injection-molded, thermo-formed and blow-molded containers. They can be printed on paper and plastic substrates, although plastics are the substrate of choice. Printing these labels on a substrate that is identical in chemical composition to the container allows for superior recycling capability.

The in-mold label is also permanent, with no delamination or image degradation. In-mold labels also add rigidity to containers, thus reducing the overall packaging weight by up to 15%. The disadvantages are cost (processing speed), and the fact that pre-labeled container inventory is increased dramatically.

## Market Growth

The growing label market, when combined with emerging customer demands, is a prime candidate for on-demand and digital color printing technology. Most of the market research quoted projects growth from 1994 to the year 1999.

The label market is a mature market and does not experience the radical kinds of changes seen in other printing and publishing markets. Over 82.5% of label orders in

1994 were in run lengths of less than 100,000. This phenomenon is forecasted to grow. Considering that many labels can be printed on a single sheet, label printing is considered a short run. In 1999 it is estimated that 89% of label runs will be less than 100,000. Average run sizes are getting smaller. Customers require just-in-time inventory and the ability to execute design changes/information updates at a greater frequency.

## Run Lengths in Label Printing

| Run Size | Market in 1994 | Estimated Market in 1999 |
|---|---|---|
| 0–25,000 | 27.5% | 34.0% |
| 25,000–50,000 | 24.0% | 29.0% |
| 50,000–75,000 | 19.0% | 19.0% |
| 75,000–100,000 | 12.0% | 7.0% |
| **Subtotal** | **82.5%** | **89.0%** |
| 100,000–150,000 | 5.0% | 3.0% |
| 150,000+ | 12.5% | 8.0% |
| | **100.0%** | **100.0%** |

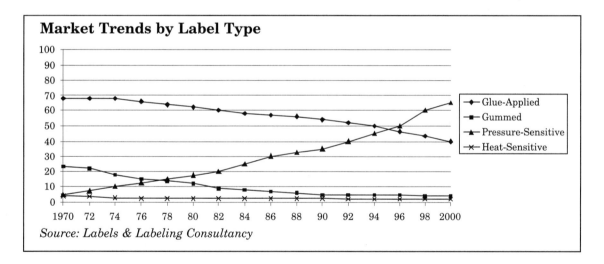

**Market Trends by Label Type**

Source: *Labels & Labeling Consultancy*

## Market Research Testing

Another motivation for short runs is the market research testing phase of label and packaging design. After the approval of the design of the package, the only way of testing the success of the new product in the market is to launch a limited number of the packaged products to the market and to examine the market response to this product and its package. In reality, there is a small ratio between products that are launched to the market and products that succeed (this ratio can exceed sometimes 15:1).

Usually, the product manufacturers are trying to change the "image" of the failed product by changing the design of the package and even the name of the product. These kind of changes derive, of course, from printing the packages all over again and throwing away the obsolete printed substrate of the old design.

The only way of launching a new product to the market today is printing the package substrate in a conventional process (e.g., flexography/gravure), and therefore the minimal quantity printed is in the order of 3,000 ft. (1,000 m).

Therefore, the throwing away of scrap printed substrates involves a substantial amount of money (the costly multilayer substrate, the work, ink, etc.), not to mention the environmental issues and costs related to the disposal of plastic substrates.

Therefore, the ultimate tool for printing relatively small number of packages (a couple of hundred meters) for examining the market response for a new launched product is an on-demand or digital color label press.

Another advantage of on-demand and digital color printing is to incorporate variable data. The advent of variable information requirements is stimulating the label market, switching packaging decoration over to labels. Currently 35% of all labels carry some form of variable information. A large percentage (38.7%) of all pressure-sensitive labels in North America are used in the personal care/cosmetics field.

The brand owners in this category change designs frequently, proliferate flavors/ variants, are aggressive in their requirements for promotional packaging, and are very image-conscious. Multiple use of labels on a single package is on the increase due to demand for greater levels of information.

Often times, growth is very rapid and hard to predict because of new FDA/Euro Standards related to nutritional information and health and safety warnings. For example, the average growth from 1987 to 1994 was 8.7%. However, when the FDA passed, FDA nutritional information legislation and the USDA passed meat/poultry information legislation, growth doubled the next year to 16.4%.

Label substrates are getting more sophisticated/expensive. The "no-look" label is gaining in popularity.

## Markets

The overall market is broken down into the *primary* pressure-sensitive/self-adhesive label category and the *secondary/identification* category. The primary pressure-sensitive/ self-adhesive label category is forecast to grow at 8.0% per annum until the year 2000.

Quality demands are changing in the premium-priced, pressure-sensitive self-adhesive category. To help fill this need, new high-tech, high-quality label systems (shrink-sleeve/labeling and in-mold labeling) are increasing in usage. This increased use is reinforcing the customer demand for premium quality/price labels.

These high-quality labels are attracting companies with premium-quality brands. For companies interested in branding, the use of high-quality labels and irregularly shaped containers is seen as another method to differentiate their products.

The secondary/identification market is growing faster. In this category, the labels have been experiencing a 20% annual growth rate for the last 10 years due to the Electronic Data Interchange (EDI) and QR (Quick Response) label market growth, i.e., bar codes. Between 1996 and 2000, the market is forecast to grow at 15% per annum.

The primary pressure-sensitive market was $2.1 billion in 1994 in North America and is expected to grow approximately 10.1% annually by year 2000 for a total of $3.4 billion in 1999. Two major market sectors, personal care and pharmaceutical and drugs, account for over 72% of this business. The personal care product market sector still continues to grow quite strongly whereas the growth in the pharmaceutical and

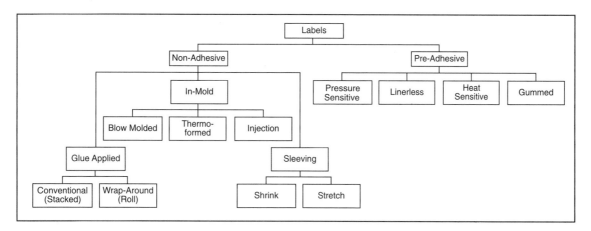

drug market sector is starting to slow down. Though small today, the beverage market sector is expected to continue to grow at a very strong rate during the next five-year period. It is expected that the beverage manufacturers will continue to promote their products in increasingly competitive markets by using PS labels to create unique attention to their products through the use of creative graphics and clear films.

**Personal Care Products.** Care products, such as shampoos, liquid soaps, deodorants, perfumes, and cosmetics account for the largest sector of the primary PS label market with sales of approximately $820 million, or 38.7% of this primary market sector.

The growth of PS primary labels for personal care products is expected to continue to expand at approximately 11% per year. At this rate, the U.S. market is expected to reach $1,377 million by 1999.

Several factors are driving the increased use of the PS labels in this market. Many personal care products are perceived and promoted to be glamorous in that the user of these products will not only become more attractive but will also lead a more fulfilling life. Therefore, the graphic image of the package projected in part through the label is extremely critical in brand selection.

Another significant factor that is driving the use of PS labels in this market sector is the environment in which these products are used. Many personal care products are used in the bathroom, which normally experiences wide variations in moisture and temperature. Therefore, wettability and long-life characteristics are necessary for the facestock materials. Also, because these packages normally must be squeezed to dispense product, the label must have good "squeezability" characteristics.

An important consideration influencing the selection of PS labels in this market sector is that personal care products normally generate a higher profit margin and therefore are more easily able to justify the higher-priced PS label.

In-mold labeling (IML) continues to compete with the PS labels for this market. IML has similar image characteristics to the PS label but one-half the cost.

But because the IML label is an integral part of the package, that package can only be used for a specific product. Therefore, a larger inventory of packages must be maintained when this labeling process is used. Conversely, the PS label can be printed and/or modified very quickly during the filling operation, thus enabling quick product changes in the production line and a lower inventory of packaging materials.

Because of the longer lead times involved, IML tends to be better used on more mature, steady-volume products. Printer/converters will continue to experience strong price pressure from this lower-cost, competitive method of labeling.

**Pharmaceutical and Drug Products.** Pharmaceutical and drug products, both over-the-counter and prescription, continue to have significant applications for PS labels. The size of this market sector was approximately $724 million in 1994, which was about 34% of the total primary PS label market.

The growth of PS primary labels for pharmaceutical and drug products is expected to continue to increase at an annual rate of about 5.8% to a market size of $961 million by 1999.

The demand for PS labels in this market segment is driven by the growing trend for brand selection by the consuming public, which in turn puts emphasis on the appeal of the packaging and thus on the label. So, graphics again is the important consideration in the use of PS labels.

In the past, most of our medication was dispensed through the local pharmacist who either followed the specific prescription outlined by the patient's doctor or who measured and tinkered with larger, bulk-like quantities of medicinal products behind his counter and dispensed what he thought was a cure based on experiences with similar ailments.

Today, the buying process is quite different. With self medication on the rise, especially for older consumers, they go to their local drug or superstore and, while walking through the many aisles of pharmaceuticals, will pick and choose most of their medications. During that moment of brand selection, the perception of the product and its ability to cure is influenced by the packaging and thus its label.

The pharmaceutical and drug industry is very competitive. More products are reaching the store shelf every day. There has been a tremendous shift from prescription drugs to over-the-counter (OTC) drugs. In addition there is a shift from brand name drugs to generic drugs, some sold via prescription and others sold OTC. Also, more regional store brands are being promoted in the various chain outlets, which is increasing the competition for the national brands.

Adding the choices of products available to the buying public, possibly to their confusion, is the decision by the FDA to speed up its approval process of new drugs. Today, more drugs are able to roll out within twelve months of requesting FDA approval. For emergency drugs, the approval time has been even reduced to six months.

**Food.** Food products accounted for approximately $210 million, or about 12% of all PS primary labels. This market segment is expected to grow at a rate of approximately 12.7% per year. By 1999 this market is expected to be about $381 million.

Where the PS label is creating inroads in this market is on foods whose sales appeal can be greatly enhanced by the graphics of a PS label.

A PS label is preferred on food products that (1) come in contact with low temperature and/or high moisture levels, (2) resist abrasion, and (3) are subjected to short runs and package changes.

Meat processors, in particular, demand a special level of service. Sudden changes in processing schedules can create a need for new labels virtually overnight. PS labels can meet these very quick production requirements.

As competition increases between national and local store brands, it is anticipated that PS labels will be demanded more often in order to enhance the competitiveness of the products involved. Also, the increasing demand by the consumer for more information about the product while it is still in its package is becoming more significant during the brand selection process.

**Beverage.** The beverage market segment accounted for approximately $150 million of the PS primary labels. This market is expected to grow at an annual rate of approximately 18% for the next few years for a market of about $350 million in 1999.

The continued use of PS labels in the beverage market segment is being driven by the need to create a certain image and/or ambiance for the beverage product. Thus, the print quality, the total graphic image, and the label appearance are key factors for the continued use of the PS label.

A current trend in the beverage market today is the "new age" look. Film facestocks are used to develop this characteristic. Beverage manufacturers feel this no-label look greatly enhances their "new age" look and increases the probability of the product being selected at the store shelf. It is known that the consumer normally takes only several seconds to decide which products and brands to put in the shopping cart. Thus, manufacturers are continually attempting to create the most positive image for their products so their brand will be selected during this very brief brand selection period.

The clear, no-label look is one of the major features accounting for the strong growth of film facestock materials in the beverage market. And, films are rapidly finding new applications in this industry.

There are over 26 billion glass bottles used to package beer, soft drinks, wine, spirits, and bottled water. Well under 1% of these bottles use PS labels. It is expected that the use of PS labels will grow substantially during the coming five years.

## Secondary PS Label Market

The 1994 secondary label market in North America was approximately $3.8 billion. It is expected to grow at 8.3% annually by the end of 1999.

The manufacturing sector accounted for the largest use of PS labels with $2,404 million, or approximately 62% of the secondary market. This sector, which consists of a great number of different PS applications, is expected to continue its robust growth.

On the other hand, the next largest market sector, the retail market which includes both food and nonfood sectors, accounted for $504 million or about 13% of the secondary market. The growth for this sector is slowing as it approaches maturity.

**Manufacturing.** The manufacturing segment is expected to grow at an annual rate of approximate 9.3% during the next few years, for a market size of $3,665 million by 1999. The primary driving force motivating the use of PS labels in the manufacturing sector is the continuing demand for manufacturers to improve the efficiency and profitability of their operations. They are accomplishing these objectives through the use of automation systems that involve the automatic identification of their products/components/packages, which in turn require PS labels.

Another driving force behind the use of pressure-sensitive labels is the need for more product information. This demand for more product information is being required by governmental regulations concerning product safety warnings, product disposal procedures, and the very recent nutritional data regulations for food products. In addition, more manufacturers are providing additional information about their products if for no other reason than as a hedge against future product liability claims.

An additional motivating force increasing the use of PS labels is the compliance requirements previously mentioned in the retail, nonfood section. As more manufacturers of retail products conform to the compliance requirements of their large customers, they are becoming more familiar with the benefits of PS labels and are incorporating them as part of their own internal control systems. Thus, new applications are being discovered for the use of PS labels almost daily.

The use of PS labels is widely spread among all the industrial market sectors. There are many common PS label applications that each industry sector has, such as inventory control, parts identification, product package shipping, and product informational labeling. Many of these applications are quite generic in that the label specifications used for shipping products in the electronics industry are quite similar to the labels used to ship products in the automotive industry. The size and graphics required might be different, but the materials used are quite similar.

## Electronic Shelf Label

A competitive technology expected to replace the use of the PS shelf label is the electronic shelf label (ESL). The ESL is designed to remotely change product prices at the shelf, thus saving labor and label material.

During the past several years, consumers and the media have been increasingly vocal about the inaccuracies of pricing in grocery stores. They have noticed that, in many instances, the price on the shelf label is not the same price as rung up at the checkout counter on the point-of-sale (POS) register. The problem is that the price changes sent to the store are not put into the pricing system in a timely fashion. It appears that performing this price change operation electronically with an ESL system will solve this growing problem.

The ESL is an electronic device, about the size of a standard credit card, that is fastened onto the same shelf edge to which the PS label is adhered. When a price change is sent to the store, it can be sent directly into the store's POS computer memory, which stores all of the prices in the store.

The ESL is electronically connected to the same database as the POS system, so that when a price change occurs, it automatically is seen at the POS checkout counter and simultaneously at the shelf edge on the ESL module. With the ESL system in place, there is no need for PS shelf labels. At this moment, about thirty grocery chains in North America are evaluating these systems, but three major chains have already completed their evaluations and are starting to purchase them for many of their stores.

During the next five years, it is expected that the use of PS labels for grocery shelf applications will experience some reduction in demand as more grocery stores purchase ESL systems. It is anticipated that, within the next ten years, they will be the primary method of shelf price labeling and replace the pressure-sensitive shelf label in most grocery stores.

## PC Board Labels

In the manufacturing of printed circuit boards, PS labels are used to identify the boards during and after manufacturing. The majority, over 80%, have bar codes printed on them for identification. The labels usually are made from polyimide and polyester depending on where in the manufacturing process the label is attached to the board.

The film is expected to grow annually at a rate of about 5–7% during the next years through 1999. Because it is quite expensive, continuing work is being done to develop a less expensive replacement for this material. Its excellent resistive qualities to temperatures and caustic materials enable it to have a market in the electronic sector, but its future is tenuous to its price.

Reliable estimates of the number of PC boards manufactured in the U.S. are difficult to find. During 1994, however, approximately 1 billion PC boards were manufactured in the U.S. About 67% of them were multilayer, 22% were double-sided, and the remaining 11% were single-sided and flex. PC boards are used to mount surface components, such as ICs and PROMs, and to connect them to each other in a predesigned circuitry.

Board production is decreasing because ICs and other surface-mounted devices (SMDs) that use board circuitry to connect them to each other are now incorporating more of these functions within themselves, thus requiring less external interconnectivity and therefore fewer boards. Not only is the production of PC boards decreasing, but the board is becoming smaller and more crowded with SMDs so that the amount of board space available to put on a PS label is getting smaller also. Industry expects that labels will continue to become smaller.

The PC board manufacturing environment is becoming less harsh as the "no clean" process is gaining acceptance. This new process uses more friendly chemicals, thus the board requires less cleaning prior to its use.

The PS label is affixed to a PC board to track its process through part of its manufacturing process and/or to identify it after it has been placed in a larger piece of electronic equipment, such as a computer. During its manufacturing process, the board often is subjected to temperatures up to 550–600°F. It also may be in contact with solder flux as well as very caustic cleaning solvents. Thus, the label may be exposed to very severe conditions. As mentioned previously, polyimide is used on the underside of a PC board that would be subjected to these high soldering temperatures. On the other hand, polyester would be used on the top side of the board when vapor phase soldering is used.

The trend is for these labels to be printed on site where the board is being manufactured. In the past, most labels were made off site by a printer/converter and shipped to wherever the boards were being made. The imaging was done using a photographic process to ensure that the labels could be read accurately and have a long life. But, with the development of thermal-transfer printing technology, the trend is for these labels to be printed on site with thermal-transfer printers. Customers are able to save not only on printing costs but also on inventory costs since they don't have to inventory as many of these very expensive labels either.

## Future Growth and Market Forecasts

Historically, the growth of labels overall has been around 3–6% per annum, and there seems little evidence to suggest that this overall level of growth will not continue.

Some types of labels are undoubtedly declining, mainly to be replaced by other types of labels or taking business away from direct decoration.

Major factors influencing the growth, or decline, of the different label methods include the trend to more and more convenience foods, the increasing use of plastics in packaging, environmental issues, the continuing need of variable information, a requirement for more promotional labels, a growing demand for form/label combinations, and the necessity for manufacturers to make use of increasingly sophisticated anti-counterfeit and anti-tamper labels. Retailers are demanding new labeling and packaging technologies to combat the very high levels of theft from stores — currently running at near 20% of turnover and about 1.8% of profits.

The trend to convenience foods — ovenable, microwaveable, instant, etc. — has perhaps had the most impact on glue-applied labeling, where the consumer now purchases less food in cans (with paper labels) and more in trays, foil, plastic, frozen, and the like. This trend to convenience foods looks set to continue with little or no potential for glue-applied labels on cans to grow.

Over the past seven or eight years, there has been a significant growth in packaging in plastics, for which the label buyers now look for pack/label compatibility for recyclability. Such a requirement has aided self-adhesive label growth — where a wide range of plastic labels are available — and has tended to have detrimental impact on direct decoration or glue-applied paper labels.

The move to packaging in plastics has also influenced the use and growth of in-mold labeling for blow-molded bottles, for injection-molded tubes, and most recently for thermo-formed containers.

The need to have plastic labels on plastic bottles containing carbonated drinks (for recycling, expansion, and clear labeling requirements) has also been of benefit to self-adhesives and, over the past two years, has led to development of reel-fed wrap-around, glue-lap labeling using plastic films. This new technology is also expected to continue growing.

Environmental issues and changing packaging waste legislation have, and still are, impacting on label usage — with both negative and positive effects. Pressures to reduce the amount of packaging, particularly secondary packaging, are leading some packaging companies and packaging buyers to look to eliminate cartons and leaflets in favor of labels. Others see labels, particularly self-adhesives, with matrix and liner waste as being environmentally unfriendly.

As yet, however, there has been little or no evidence of environmental issues or packaging waste legislation having any significant impact on the use of labels.

Probably the most important influence bringing continuing rapid growth to labels generally and to pressure-sensitive labels specifically is variable-information printing. The nonfood retail sector continues to look at and invest in electronic point-of-sale (EPoS) and the use of bar coding with variable text, graphics, and pricing. This is creating new label opportunities for self-adhesive labels in areas such as hardware, garden centers, clothing, pharmacies, plumbing, car care, electrical, etc.

Most of these sectors are expected to become significant users of labels for carrying bar codes and variable/fixed information. If they also follow the pattern of variable information growth in more traditional supermarket labeling, then the labels will start as relatively simple, one- or two-color labels, and then gradually be printed in more and more colors, become larger, carry increasing amounts of both fixed and variable data and start using promotional and coupon labels at an increasing rate.

Although not yet particularly large in size, the form/label combination market has already become important and continues to grow in excess of 12–14% per annum. Much of this new market, however, is being attained by the business forms printer (who understands the computer imprinting market and already sells forms to it) rather than the label printer.

With worldwide losses from counterfeiting of goods — including the packaging and labeling of such products — now running at an estimated $150 billion a year, the pressure to come up with increasingly sophisticated anti-counterfeit label technologies is now creating a whole new sector of the label industry — a majority of which focuses on the self-adhesive market.

Likewise, retail theft is now estimated to bring annual worldwide losses in the region of $90 billion per annum. Major retailers are now saying that they would like all items sold retail to be protected by anti-theft labels or packaging by the end of the century. The potential for the label industry to solve or minimize the problem of retail theft is enormous.

Already in the pipeline and being successfully trialed are new generations of supertags and superlabels that carry encodable information and that enable a whole shopping cart of goods to be scanned without removing each item from the cart. One scan enables a register receipt itemizing all goods in the cart, together with appropriate prices, to be produced. Stores are already looking to the day when they can introduce such technology.

Labels embedded with electromagnetic, radio-frequency, or microchip technology are also now coming to the market and are expected to find increasing usage in new label applications. Such "super" labels or "intelligent" labels are seen in new label applications, new methods, and new label materials and technology, and they are expected to provide the total label industry with sustained growth to at least the end of the century — and probably well beyond.

Best placed to benefit from the new opportunities and markets is the self-adhesive label, which now has an underlying growth trend in almost all developed world markets of around 10–12% per annum. Major world players in the self-adhesive market see the underlying growth trend for self-adhesives growing to perhaps 15% per annum by the end of the 1990s.

Market share forecasts for the main labeling processes over the next few years and up to the year 2000 are shown in the following table.

## Market Share of Main Labeling Processes by 2000

| Type of label | Market share |
|---|---|
| Self adhesive | 56.3% |
| Glue applied | 32.6% |
| Shrink/stretch sleeve | 6.1% |
| Gummed | 2.4% |
| In-mold | 2.6% |
| | **100.0%** |

Even the new growth technologies of shrink/stretch sleeving and in-mold labels between them are only growing at around one-fifth the rate of self-adhesives, and for the foreseeable future, they will not impinge on self-adhesive market share.

## Products

Many of the devices included in this book can be used for printing labels. Two of the manufacturers have made special enhancements for packaging and label products: the Omnius by Indigo and the Nilpeter DL-3000, a Xeikon-based DCP/32S from Nilpeter.

The Nilpeter DL-3000 was shown for the first time at Labelexpo in 1996. It is a single-sided label printing system with a choice of front ends and finishing options. It is currently in testing trials in Europe and is expected to be available for purchase in 1997.

The Indigo Omnius web-fed digital press was introduced in early 1995 and has been successfully installed in the operations of several major customers around the world. During the past year Indigo has formed partnerships with leading label substrate manufacturers, with the result that 39 different substrates are now available optimized for Indigo's proprietary ElectroInk liquid-ink technology.

---

### Acronyms

| | | | | | |
|---|---|---|---|---|---|
| EDI | Electronic data interchange | IML | In-mold labeling | POS | Point of sale |
| EPoS | Electronic point of sale | OTC | Over the counter | QR | Quick response |
| ESL | Electronic shelf label | PS | Pressure-sensitive | SMD | Surface-mounted device |

---

| Term | Definitions/Description |
|---|---|
| Label | Pre-printed layer of material (paper/plastic/foil) attached to or accompanying an article and which provides appropriate information. Labels can be subdivided into *primary* and *secondary* labels. |
| Primary labels | Labels that are designed to attract attention and contain information to appeal to a prospective buyer or user of the product or contents of the carton. These labels are the most sophisticated from a substrate perspective and feature the most ornate designs. Image representation is their primary feature. |
| Secondary labels | Labels that carry additional information such as nutritional details, instructions for use, cautions, warnings, pricing, couponing, or special deal information. While printed from similar substrates to primary labels, they are made more austere, often printed in one color only. High utilization of variable information printing in this category. |
| Pressure-sensitive (PS) labels | The premium end of the label market. These labels are manufactured from the most (primary label) sophisticated materials. PS labels are self-adhesive on which the adhesive is active or ready for mixed rate application. The labels are pre-diecut and are individually attached to a silicone-treated liner or "release agent." Easiest type of label to apply. The fastest growing market sector. |

| Term | Definitions/Description |
|---|---|
| Glue-applied labels | Most common label type. Preprinted, particularly on paper, they are often referred to as "cut (primary label) and stack" labels, as they are cut to size after printing. Glue-applied labels can also be supplied in roll format. Labels have glue applied to them just prior to application to the surface of the container. Dominant form of labeling in food (can) market. Typically glue-applied labels are coated with protective coating. High-volume labeling characterizes this application. Low cost, low quality. The commodity end of the business. |
| Heat-seal labels | Sometimes referred to as thermoplastic labels. They are preprinted on various laminates (paper/foil, paper/polypropylene) that have been coated with a latent adhesive. The adhesive remains dormant until heat-activated to a tack level necessary to promote bonding to the container. These labels are often supplied "stacked" but are also available in roll form. |
| In-mold labels | Labels that are permanently bonded to a container during its manufacturing process. These labels are available for injection-molded, thermo-formed and blow-molded containers. They can be printed on paper and plastic substrates, although plastics are the substrate of choice. Printing these labels on substrates that are identical in chemical composition to the container allows for superior recycling capability. The label is also permanent with no delamination or image degradation. In-mold labels also add rigidity to containers, thus reducing the weight of packaging materials by up to 15%. Disadvantages are cost (processing speed) and the fact that prelabeled container inventory is increased dramatically. |
| Heat transfer labels | These labels are used when a high degree of squeezing or flexing occurs, such as in plastic tubes and bottles. These are similar to decals insofar as the image is not printed on a normal substrate. The images are printed onto a lacquered film that has been printed on a carrier web pre-coated with a release coating. Application of heat by a platen press (iron) onto the ink printed area of the web generates the transfer of the image. This is an expensive but high-quality labeling system. This type of heat transfer labeling is prevalent in the cosmetics industry. |
| Shrink-sleeve labels | These plastic films are reverse-printed and applied by heat (shrink sleeve) or by stretching or pressure (shrink labels). They can be individual (sleeves) or in the form of a roll (label). The substrate of choice for shrink labels is polypropylene, while for shrink sleeving, PVC or polypropylene are the substrates of choice. This is becoming a more common labeling system favored by the beverage (soft drink), household chemical, and health and beauty sectors. Shrink labels are for high-volume applications, while shrink sleeves can be used for unusual/irregular shaped containers. |

# Section II

# New Issues

# Chapter 10

# Marketing and Selling On-Demand Services

It is pretty easy these days to get caught up in the latest technology. It is on the front page of the newspaper once a week and the cover story of one national magazine at least once a month. But as most experienced CEOs and managers know, there is more to selling a successful product or service than technology alone.

That is what this section focuses on, the factors that influence the successful delivery of on-demand products and services that are not related to printing technology. Some are indirectly related to the printing process, such as product positioning, marketing, and variable-printing direct marketing pieces, while others are more directly related to printing, such as issues with paper and the bindery.

## TV Programming and ATM Cards

The chapter on market research discusses a theory of how technology changes consumer demand and/or consumption. We discussed how inconceivable it would be to open a checking account today if that bank didn't offer an ATM card and how TV changed the programming of radio shows. In essence, these new technologies changed customer demand. We concluded that it is possible that on-demand printing solutions will change consumer demand.

Here's a riddle: *How are TV advertising, banks that offer ATM cards, and on-demand printing related?* Think about it for a minute: TV advertising, banks that offer ATM cards, and on-demand printing.

You could say that ATM cards and on-demand printing are fairly new, but not TV advertising. You might think that TV ads are expensive, but ATM cards are not.

Give up? The answer is that all three provided a service in a new and different way, based on new and different technology, and most importantly all require unique product positioning.

You would not want to describe the advantages of TV advertising in the same terms as radio advertising. You would not promote the advantages of checking with an ATM card the same way as a checking account without ATM availability. And you should not position the advantages of on-demand printing the same as the advantages of traditional printing.

Why? For two reasons. First new technologies typically have additional costs associated with them. Therefore, you don't want to compare the new, more-expensive method to the older, less-expensive methods. It would be difficult to sell that new service.

Clearly, it costs more to produce a TV show than a radio show. If you are a bank, it costs you more to have ATM access all over the world than to only have local-branch availability. And, as we will discuss in several chapters, it may cost you more to offer on-demand services because of equipment costs, paper, consumables, variable printing, connectivity (network), and bindery issues.

Second, the advantages of products and services based on new technology are that they are more valuable to the client. TV ads are much more valuable than radio spots. Banks that offer accounts with ATM availability are much more valuable than banks without them. Ultimately, printers and service providers that offer on-demand services will be more valuable and effective than those that don't.

Another phrase for "much more valuable and effective" is value-added or developing customer intimacy. Value-added is a differentiating factor that increases perceived value or competitive value of the company.

## Value-Added

Value-added strategies are obvious in all markets. According to market share, people prefer Federal Express because of the perception of hassle-free service. Parents run to buy Disney theme park tickets to try to catch a piece of "the magic." Neither of these products or services are less expensive than the competition. In fact, they cost more. How do they achieve a greater market share? It is based on their value-added or perceived value perception.

This even occurs in the home repair market. Why are Home Depot's competitors losing market share to this fast-growing retailer of do-it-yourself supplies when they are all selling similar goods? It's not because their products or advertising are significantly better than the competition; it's because they work closer or more intimately with their customers.

Customer-intimate companies are willing to spend time and money today to build customer loyalty for the long term. They typically look at the customer's lifetime value to the company, not the value of any single transaction. This is why employees in these companies will do almost anything to make sure that each customer gets exactly what he or she really wants. Nordstrom is one example of such a company; IBM in its heyday was another; Home Depot is a third.

Home Depot clerks spend whatever time is required with a customer to figure out which product will solve his or her home-repair problem. The company's store personnel are not in a hurry. Their first priority is to make sure the customer gets the right product, whether its retail price comes to $59 or 59¢.

Individual or customized service is Home Depot's forte. Clerks do not spend time with customers just to be nice. They do so because the company's business strategy is built not just around selling home-repair and improvement items inexpensively but also building long-term relationships and servicing the customer's needs for information and service.

Some companies know their customers so well that they can predict when their purchases are likely to slow or increase. Some go so far as to prioritize the value or importance of their customers, based on profitability or sales volume.

These companies are practicing relationship marketing — a technique driven by knowing what appeals to your customers. The results speak for themselves — the more successful you are at understanding and building a relationship with your customers, the more successful you will be.

What is the first step in understanding your customer's needs and working closer with your customers to develop greater value-added? The first step is understanding what the advantages of the new technology are compared to the old technology.

## The Four Advantages of On-Demand

There are four main differentiating factors of on-demand products and services:
- Reduced cycle time
- Shorter run length
- Cost per copy of short runs
- Variable printing

For certain market segments, "time is money" more than "money is money." For these market segments, electrophotographic or toner-based on-demand printing technologies offer significant advantages. It is possible to walk into a facility using an on-demand press at 9:00 AM and walk out by lunch time with 50 or 500 impressions.

When the on-demand presses where first introduced into the marketplace, the run lengths were estimated to be between 35 and 5,000 impressions. Since then newer technologies have been introduced that compressed that market. On the high-volume end of the spectrum are the direct-to-plate-on-press options (Heidelberg Quickmaster, Omni-Adast) and plate and press enhancements such as automatic plate mounting.

On the shorter side of the market, the technological advancement shifting the minimum number higher is faster color copiers. As a result, the appropriate market today starts at about 50 impressions. Anything shorter could be done on a high-speed copier. On the higher end, the market has been reduced to 3,500, but most shops do not exceed about 1,000.

One company with experience with several on-demand technologies is Linotext America Inc., which has three Heidelberg GTO-DIs and two AM Multigraphics' Xcikon DCP-1s. Housed within a 20,000-sq.ft. warehouse, it is perhaps the largest digital printing facility in the world. Former Stanford University engineer Duane Sincerbox works for Linotext and says that although the GTO-DIs can handle runs of up to about 30,000, most jobs average between 1,000 and 2,500. In contrast the Xeikon runs are shorter, between 25 and 1,500, averaging 500.

Cost is another differentiating factor. Imagine that you had to print one page off a press and had to stop that press and sell it, that one page could cost $10,000. If we print two pages they would cost $5,000/each, and three would cost $3,333/each. As the print run increases, the price per page decreases.

Copy technology on the other hand is different. If it costs $0.45 a page for the first page, it will cost the same for the second and third pages. If a client is paying $3.00

a page, copy-based technology is less expensive than traditional printing technology until a certain break-even point is reached.

As we discuss throughout the book, one of the strongest selling points of on-demand products and services is the ability to use variable information to customize and personalize the product and increase their sales effectiveness.

A question that we have posed to many market research firms is how much variable printing is actually done. Since we have not discovered any reasonable methods of analysis, we estimate that only a small portion (about 1%) of the overall printing done in the U.S. utilizes this advantage. We are not really sure why variable printing is not catching on, but one possible reason is that the various software (i.e., database software) packages are fairly new.

## The Five Selling Factors

In the last section we discussed the importance of understanding the benefits of the technology for the customer. The next consideration in the successful sales of on-demand products and services includes:
- Product positioning
- Promoting the benefits
- Reconsidering your contact
- Thinking creatively
- Overcoming buyer resistance

**1. Don't position it as the same old thing.** There was an effective advertising campaign a few years ago used to sell cars. It said "This is not your father's car." It was an effective selling tool because it drew a clear picture in the minds of the customers. This picture helped position and differentiate the purchase in the minds of the customers.

Positioning is important when selling on-demand services. Customers need to know the differences, the advantages, the unique impact, and potential "breakthrough results" that on-demand products and services have to offer. Because in the end, if on-demand printing doesn't do a job better, faster, or cheaper, what's the use?

**2. Learn and promote the benefits.** To extend the analogy one more step, service providers need to present their "This is not your father's car" message to their market — clearly, distinctively, and efficiently.

A common mistake made by many new on-demand service providers is to send out a traditional marketing piece by mail, touting their digital press's specifications. Although it is important to "get the word out," there are two problems with this approach. First, a traditional printed piece does not demonstrate the power of the technology, and second it may not discuss the advantages.

A traditional printed piece would not take advantage of one of the greatest strengths of on-demand printing — the variable information aspect. Adding variable data would clearly demonstrate the power of personalization. Think about your own experience. You get two pieces of direct marketing in the mail. One says "Congratulations, you've won $1 million," and the other says "Congratulations Howie Fenton, you've won $1 million." Which would you open first? Well, if you were Howie Fenton, you'd most likely open the one with your name on it.

Therefore a simple variable information solution would be to personalize each piece with the customer's name. A more elaborate personalization might include the customer's picture.

Second, a traditional marketing promotion usually discusses equipment specifications; e.g., how may pages per minute or impressions per hour, the maximum line screen, or different screening technologies.

Although interesting in a limited sense, this information does not discuss the unique advantages of the technology. More importantly it does not discuss how the technology will help them to do their job better. Here's a few do's and don'ts:

- Don't just say that the equipment prints directly from diskette. Describe how the print runs can be shorter, for less money, and allow for more frequent updates.
- Don't describe the dot per inch, lines per inch, or levels of gray per pixel. Instead, discuss how a more targeted piece will cost less and result in a higher response rate.
- Forget about the fact that graphically shows how many steps have been eliminated. Sell the fact that, because of this, they can walk in the door at 8:00 AM with a job and walk out at lunch time with printed sheets.

Some good salespeople have the ability to identify and exploit a customer's "hot buttons." This is one example of generating excitement in the customer. If you can find a way to get the customer excited, then they will sell themselves on the company. The key is to show the customer how the technology can help them meet their goals.

We know that what we are about to say is hard to believe, but if you can demonstrate how the on-demand technology is more effective, then the cost issue will become less important. For example, if you increase sales from 2,000 products to 6,000 products, who cares if it increases the costs from $0.25 to $0.45 each?

A good example is K&S PhotoGraphics, a Chromapress site in Chicago. They do test marketing mailings of sample designs to thousands of potential customers and solicit their input. In the past, these samples were printed using four-color offset, or printed on individual 8×10-in. sheets of photographic paper by K&S — a commercial photo lab. Now K&S outputs the sample designs on a double-sided, 11×25.5-in. Chromapress sheet, which is folded into a convenient, six-page booklet for mailing to prospects.

**3. Reconsider your contact.** Many early users discovered that the person they sold traditional services to was the wrong person to sell on-demand services to. Cardinal Communications Group in New York City discovered this after they installed two Indigo E-Prints in 1994. According to Val DiGiacinto, the senior vice president and chief operating officer, "Our marketing initially was not geared toward the right person. It's not the print buyer we need to educate about digital printing. By the time the buyer receives management's request for a print order, it's too late . . . the order is far too large a quantity. We realized that we had to get to the higher levels of the organization; we needed to market to the marketers."

"These are the people who make the printing decisions, the ones who truly can understand the marketing potential of this technology," DiGiacinto continues. "We must show them that the 50,000-run job they were going to print can be far more effective if produced in targeted, more personalized lots of 500, for example. Buyers have never ordered printing in such small quantities before, so getting them to rethink their habits is a real challenge."

Others believe that it is the creative staff who should be the new contact. According to Sanjay Sakhuja, the president of Digital Prepress, Inc. (DPI), a Chromapress site in San Francisco, "Marketing digital work is not in the same league as marketing traditional offset printing. Printers must get involved at the conceptual stage of a project, literally helping a client create a product from nothing. Most people in our industry are not used to thinking in those terms, but it's a task we took on as part of getting involved in this technology."

Linotext Digital Imaging, another California-based printer, used a very subtle technique to "get the word out" to designers in the area about the purchase of its Heidelberg GTO-DI. Instead of a traditional direct mail campaign Linotext sponsored a contest encouraging customers to design a logo for its digital printing services.

**4. Think creativity.** The companies experiencing the greatest success are those that understand the advantages and figure out creative ways to use these advantages to benefit their customers. Let's look at a few clever ways to sell this service and start with some interesting uses of digital cameras and variable data.

A printer in Southern California sets up a digital camera in automobile showrooms. When a potential buyer sits in the car to get a feel for it, the camera takes a picture. One week later the dealership sends that potential buyer a follow-up brochure. This is not a traditional brochure, but rather a customized and personalized brochure with the picture of the person sitting in the car he or she admired, perhaps even in the color they desired.

As a child sits on Santa's lap to discuss his or her Christmas list, a digital camera captures the moment. Immediately the picture is incorporated into a predefined Christmas card template and printed from a color printer. As the parents gather up the child and walk away, they are handed the picture and asked if they would like some customized Christmas cards. Another company has created Christmas storybooks and uses digital photos of a specific child as the star of the action.

A service provider has an interesting idea. They work with a pizza company and create a promotion for free pizza for kids on their birthdays. The service provider creates the promotion for a moderate price and pays the cost for each free pizza.

When the youngsters come in, they don't have to pay anything, only agree to have their picture taken and give their name and address. The service provider uses the list to create customized and personalized advertisements for local merchants. One year the pizza kids get ads from a bicycle shop, another year from a computer shop, and one year from a car distributor. The best part is that all of the customized pieces are created by the original service provider.

A large company sends sales staff on trips to reward them for achieving outstanding results. The printer who services this customer offers to create personalized phone cards to be placed in their welcome packets.

A company uses its digital press to create customized catalogs that are geared toward a client's previous purchases. When a person buys an audio component from a stereo store, the store will follow up with a catalog mailing. This is not an ordinary catalog, but rather one with a cover that features a picture of the component that the specific customer purchased, surrounded by a series of other related components that would complement that sound systems component.

Someone at DPI (Digital Prepress, Inc.) in San Francisco notices that real estate promotions about homes and apartments appear amateurish because they are created by typing on letterhead and cutting and pasting pictures. Scanning the photographs, recreating the letterhead, and adding a little personalization by adding specific realtors, they were able to create better looking pieces that raised the bar for the market in that area. The same company has developed a product for engineering and architectural firm portfolios to replace often messy cut-and-paste jobs and for modeling agencies.

Hollis Digital Imaging (Tucson, Ariz.) uses a digital camera to shoot pictures of local high school sports teams. It designs and incorporates these pictures into professional-looking trading cards (e.g., baseball cards) to be sold by the individual school booster clubs. They find a corporate sponsor to underwrite half the production cost, and charge the balance to the booster clubs. In addition, they have created posters. Everyone has gone wild over the idea, and the company has signed up over 30 schools for the service.

Following up on their baseball card success, they are testing automobile cards and home moving cards. The auto cards allow automobile enthusiasts to have cards of their prized autos. The moving card has a picture of the new home and address, which is mailed to friends and family as a postcard.

**5. Overcoming buyer resistance.** One of the strongest motivations of on-demand printing can be a problem for first-time, non-risk-taking buyers. As discussed earlier, one of the greatest advantages is reduced cycle time or fast turnaround, such as walking in the door at 9:00 AM and walking out at lunch time with 350 copies.

But, what happens if you're in an emergency, quick-turnaround situation and you've never had it produced on-demand before. For some, it can appear intimidating and high risk. Even those not under deadline pressure often are skeptical about the turnaround and quality.

There is no doubt that buyers must be comfortable with the procedure before they use it. Different companies use different approaches to overcome these concerns. Some offer tours and let customers watch production and talk to staff, and others offer seminars.

Hollis Digital Imaging holds regular educational seminars with prospective clients to familiarize them with the process, but these are no ordinary seminars with people sitting in a room listening to a lecture or watching a canned demo. These are interactive. According to Bob Diehl of Hollis, "Prior to the seminars, we send attendees a template for a postcard design. Clients then design the postcard using their own files and use the result as admission to the seminar. We then print the card on the Indigo so they can see the results first-hand. It's a great marketing technique and gives people confidence that the process can work wonders."

At the VuePoint 96 show, Tom LaVigne, president of LaVigne Press, said the biggest challenge to selling on-demand printing was the need to educate potential customers about the realities of the technology. His company responded by holding a number of seminars to show what could and couldn't be done.

LaVigne said that it is an ongoing battle to get the idea out of people's minds that they needed to purchase at least 1,000 copies of any job. He said that regardless of what a customer asks for, he always quotes jobs at a number of quantities, all the way down to 50 copies. He said that people also have the tendency to say no to a proposal on the basis of the fact that it breaks down to $40 per copy, without considering if a total bill of $400 or $500 is worth paying to address the need.

# Chapter 11

# The Role of Paper

*Riddle:*
*What factor slowed the availability of printed products immediately after the invention of the first printing press as well as after the digital presses shipped?*

*Answer:*
*The availability and cost of paper.*

## Circa 300 Years Ago

For 300 years after the invention of Gutenberg's press, most publications were out of the reach of the average European. The problem was not press technology but rather the high cost of paper, which, until the end of the 18th century, was made by hand from the pulp of linen rags. It wasn't until the 1790s that a clerk named Nicholas-Louis Robert recognized that there was a future in cheaper paper and left his position in a Paris publishing house to learn the art of papermaking.

Several obstacles stood in his path. He had to leave the papermill at the great paper center of Essonnes because he didn't get along with the employees. But he persisted and created a prototype machine. But it wasn't commercially acceptable. Bryan Donkin, an English engineer, took Robert's idea and turned it into the first practical papermaking machine. Ultimately the papermaking machine took off throughout Europe, Russia, and America, and the 19th century was being called the "Age of Paper."

## Circa Three Years Ago

After the introduction of the Indigo presses, an erasability problem was discovered. A special paper treatment was subsequently developed. At first, treated paper was only available in limited quantity and on limited stocks. Some companies frustrated by the limited availability bought the pretreatment machine themselves.

After the early Xeikon-based machines were delivered, only a limited number of suitable paper stocks were available. Many of the paper stocks are made in Europe and manufactured for the European-designed Xeikon engine. Some U.S.-based users of Xeikon-based machines had to buy these huge rolls of paper and get them shipped to the United States. This was not a good solution.

Shipping the rolls was expensive, and the delivery times were long. When they arrive, they have to be rerolled to fit on a common U.S. roll size. These factors fly in direct conflict with the selling points of on-demand printing: fast turnaround and low cost.

The moral of the story is that printing technology alone is not enough. Successful printing also requires the availability of low-cost paper.

## Application-Specific

It doesn't matter if you are printing on an office copier, a digital press, or a high-speed printer, there are issues with paper that can affect the quality and productivity as well as your ability to get it done on time and within the budget.

The specific technology and equipment determines the paper, it's availability, cost, and runnability. In on-demand printing, there are two general paper markets. One is the well-established, mature copier/printer market. The other is the new, relatively immature, electrophotographic-based digital press market. For the well-established copier-based products, the issues are not with availability but rather cost and quality.

In contrast, the issues for the paper for on-demand presses center on availability, runnability, price, and delivery time. Is the paper available locally, nationally, or internationally, and how long will it take to get it? Will the paper run untreated or does it need a pretreatment? Will the high temperature of the fuser cause inconsistent toner adhesion or make the paper discolor or blister? Can it be bought in small lots or large rolls? Are there added costs for pretreatment or size conversion?

## Objectives and Markets

Before addressing these questions, we need to reiterate the objectives, identify the markets, and review some paper basics. The objectives of on-demand include printing done on short notice, printing done in shorter runs, printing that satisfies the customers' needs, and printing that is efficient and profitable. If you can't obtain cost-efficient small quantities of paper or if the paper needs expensive pretreatment, then you can't sell the jobs in a cost-efficient, profitable manner. If you can't get the paper quickly, then you can't do the job on short notice. If the ink peels off the paper or the paper blisters, you may not satisfy the customer. Therefore, to compete successfully with on-demand printing, you have to understand the issues and develop strategies to overcome the obstacles.

Where are the markets? According to Charles A. Pesko Ventures (CAPV), the vast majority of paper consumed in the on-demand market today is used by office copiers and the high-speed black-and-white copier/printers such as the Xerox DocuTechs. In 1995, paper used in on-demand applications was 650,000 tons including copiers and 250,000 tons excluding copiers. Of this 250,000 tons, only 12,000 were used by four-color devices.

But the usage in the market is shifting, and future growth will be in the four-color category. CAPV predicts that while the 1995 four-color on-demand market is 2.6%, by 1999 it will grow to 4.2%, or 235,000 tons.

According to Bob Hieronymus, senior marketing project manager for Georgia-Pacific, "Forecasts indicate that by 1998 approximately 100,000 tons of paper will be sold for digital presses annually. Now, we're waiting for the market to tell us what it needs."

## Electrophotographic Papers

Despite the dreams of the paperless office in the 1980s and the infatuation with the Internet in the 1990s, the fact remains that paper is the overwhelming tool of choice for business communication. It may be copy machines, laser printers, or fax machines — all of these office machines are available as electrophotographic devices.

Paper quality influences the quality and productivity of electrophotographic printing. In paper terms, this is known as the *printability* or the quality of the image as printed, and *runnability* or how well the paper performs in the printing or copying process.

Choosing the right paper for those devices is critical. The right paper ensures accuracy, reliability, and permanence. The right paper is designed to meet manufacturer's standards for printability, readability, toner retention, moisture content, and strength. And lastly, the right paper should not jam or misfeed. Here are a few criteria to consider before choosing a paper for electrophotographic engines.

**Weight.** Weight affects a paper's durability and its ability to move through the print engine. Other important attributes that also help avoid jams are accurate trimming to proper size, proper surface characteristics, stiffness, and the ability to withstand intense heat.

Paper is often identified by its basis weight. *Basis weight* or *substance* is the weight versus its surface area. For example, standard copy paper is known as "20 pound" or "sub(stance) 20" bond. Most popular papers are bought and sold at a number of dollars per ton, or per hundred weight, or per pound. The basis weight is the weight in pounds of one ream (500 sheets) cut to the grade's basic size.

Every paper grade has its own typical basic size. The basic size of bond paper is 17×22 in. A ream of 500 sheets of 20-lb. bond paper at its basic size weights 20 lb. These reams will be cut into four 5-lb. reams of 8.5×11-in., 20-lb. bond paper used for office copiers. Similarly a ream of 24-lb. bond paper in its basic size weights 24 lb.

**Toner adherence.** Toner adherence is a critical factor in electrophotographic-based printing. The three factors that determine toner adherence are:
- Paper curl
- Moisture content
- Electrical properties (conductivity and resistivity)

Laser printers, LED printers, and xerographic printers work via electrophotographic or electrical charge transfer. If the moisture is too high, the electrical charge is dissipated resulting in uneven transfer to paper.

Several factors can affect the paper curl. The high temperatures (300–400°F) in high-speed printers and copiers can induce curl.

Slower desktop printers and copiers use a print engine whose internal temperature is lower, perhaps 150°F, which does not cause as much of a curling problem. Instead of a curl problem, however, these slower-speed engines allow the paper to stay in contact with the engine longer. This contact time is known as dwell time, which is another factor that contributes to paper curl. Papers with heavier basis weight curl less.

Users sometimes want to run preprinted papers, such as letterhead, through electrophotographic devices. Certain kinds of printing such as thermography (raised printing that simulates engraving) can damage a printer or copier. If you need to use preprinted materials, you should talk to the printer about using thermal-resistant inks and fast packaging to protect against edge damage and moisture penetration or loss.

If thermal-resistant inks are not used, ink may eventually come off the drum or fuser roll of the print engine. If the paper package is too tight, it may induce a curl. With any new paper, it is a good idea to do some short-run testing before committing to a long run.

**Other criteria for choosing a paper.** Other factors that influence output include smoothness, opacity, brightness, long-term storage (archival), and recycled content. For paper to accept toner best, it needs a level surface and a fine smooth finish. This is most obvious with flat tints of pictures (halftones) that contain fine screens or subtle tones.

Opacity affects a document's legibility and is determined by the thickness of the paper. A paper with high opacity allows no "show-through" from the other side, which is required for two-sided imaging.

Similar to opacity, brightness affects readability as well as how well the images "pop" off the page. The brighter the paper, the more legible the type and the greater the contrast of the graphics.

For archival documents (documents that are stored for extended periods of time), special paper is required. These specially made acid-free papers can extend the document's life to 200 years or more.

## Xeikon Engine Paper Issues

As described earlier, much of the original papers for the Xeikon-based products were manufactured in Europe in European sizes. To use it required that the paper be converted for American presses.

Some Xeikon users had to buy the entire mill run of a particular paper and then have it converted to the proper roll size. This resulted in a number of unexpected issues. Often the shipping became an additional expense because the roll is 40,000 lb., and after using the portion needed, the printer has to store it, which also has an expense and risk associated with it.

Finding suitable coated stocks is also a challenge. During the fusing process, the toner and paper are exposed to some very high temperatures known as radiant heat. This heat could affect the glossy appearance, remove the moisture, and transform the paper into a brittle "potato chip"-like substance, affect the paper's ability to accept toner, or make the paper unusable.

"Printers don't have the same flexibility with coated stocks as they do with uncoated," points out Glenn Toole, director of marketing digital imaging for AM Multigraphics. "While some coated papers withstand high temperatures, others produce a less glossy image." According to Bruce Iannatuono, president of Chesapeake Printing (Baltimore), a Xeikon user, "It is difficult to maintain toner concentration with coated paper. The fusion process blisters the paper, so we can't apply as much ink to coated as uncoated stock. Also, with coated and uncoated, the intense radiant heat removes excess moisture, which makes the paper very brittle. It is so dry that it cracks like a potato chip when folded."

Fortunately, more suppliers in the United States are stocking Xeikon paper, but finding a local distributor may be difficult. If you can't find a local supplier, buying long distance will slow delivery and increase costs.

## Qualified Xeikon Paper

Not all papers print well with the Xeikon-based engine. For example, using lighter-weight papers may not work well because of show-through problems. Smooth papers tend to work better than textured papers because the toner may not adhere to the valleys as well as it does to the peaks. Care should be taken in running pressure-sensitive labels. You wouldn't want to melt the backing material.

How do you know if a paper will work well in your Xeikon, Chromapress, or IBM InfoColor 70 (IBM 3170)? Digital press manufacturers, resalers, and distributors have begun a "qualified" product analysis. They test selected conventional offset papers on their devices to evaluate the material.

According to David Getlen, Chromapress product manager for Agfa in Ridgefield Park, N.J., "In the qualifying process, we establish optimum conditions, such as image quality and working parameters. For example, we determine whether the paper can run at a wide range of fusing temperatures, how much product flexibility it offers the user, if image density can be adjusted, and more."

As the market for "qualified papers" grows, more and more paper manufacturers express interest and become authorized dealers. Mohawk Paper Mills' Options, for one, is an uncoated, multipurpose paper qualified for the Xeikon as well as other presses. Introduced this past fall, it is laser-guaranteed and said to eliminate blistering and ink holdout problems.

Russell Field offers nine coated and uncoated papers qualified for the Xeikon engine, including 4CC, an industry-standard, bright-white paper with a smooth finish, for which it is the sole North American distributor. Microprint DCP from Georgia-Pacific, a paper designed for digital color presses, is qualified by Xeikon for its engines. The bright, smooth substrate is available in 60-, 81- and 110-lb. basis weights.

## Indigo ElectroInk Issues

For many people new to on-demand or electrophotographic-based (copy) printing, the critical role of paper did not become obvious until Mike Bruno, a well-known printing industry consultant, found that he could erase the Indigo ElectroInk (ink). As described in the Indigo chapter, the ElectroInk has to dry very quickly. When transferred to the blanket and heated, it acts more like a film or plastic, drying on top of the paper, than an ink that binds to the paper. This was the first indication of the unique way that on-demand technologies work with paper.

The Indigo ElectroInk is a radical departure from the first-generation liquid toners used in copiers the 1970s. The first-generation liquid-toner devices used a highly flammable solvent carrier to fuse the ink to the paper. The ink was not erasable, but the solvent was highly volatile.

ElectroInk itself has gone through several variations. It was originally licensed to the DuPont/Xerox joint venture DX Imaging for its never-marketed color proofing and duplicating process. After that the technology was tried in a product called the Electro-Press, a product that was withdrawn from the market when AM Graphics experienced financial difficulties.

However, the technology did prove successful in Israel where the E-Print was developed. According to the "erasability" story circulating in the industry, Indigo was not aware of this problem because it did not exist in Israel. According to the story, either the papers available in the Middle East or the dry environment allowed the ElectroInk to adhere more permanently.

Although optimum printing on the Indigo is not guaranteed without the Sapphire treatment, some stocks do run acceptably without the pretreatment, reports Judy Finlay, product marketing specialist, paper and substrates, for Indigo.

**Indigo Sapphire Pretreatment**
As we discuss in the Indigo chapter, a solution has been created to increase the adherence of the ink to paper and overcome the erasability issue. This process, called "Sapphire," is a pretreatment of the paper with a chemical. Our understanding is that the process involves adding a special chemical component to the paper, and that this chemical can be added by running the paper through any A3 or larger single-color printing press.

This process is similar to running the paper through a press with the dampening system on. This result is that the paper fibers rise off the paper, enabling the Electro-Ink "film" to adhere to the fibers. Although the erasability problem has been solved, new questions have emerged regarding who will apply the coating, how long will it take to apply to the paper, and how will it affect paper cost and turnaround time.

When the Sapphire treatment first became available, it was difficult to get. For one pioneering Indigo user, Speed Graphics, this was unacceptable. With an unfilled customer need and problems with getting treated paper, the New York City-based company decided to buy the Sapphire coater.

Today Speed Graphics treats paper rarely, for two reasons. First, treated paper has become more available. Second, the pretreatment process is laborious because individual sheets have to be fed individually through twice.

Similar to the Xeikon paper availability issue, the availability of Indigo papers is improving. More and more papermills are going through the approval process with Indigo to receive approval to become an Indigo paper treatment facility. S.D. Warren, for example, has recently been approved for the Sapphire treatment.

Papermakers also are finding ways to combat the slow, cumbersome distribution process. For instance, Russell Field has added new distribution sites in Reno, Nev., and Baltimore, Md., to provide next-day service. Clearly for companies to meet the objectives of on-demand printing, they need to be able to access the paper in a few days.

**Presstek-Based Digital Presses**
The two Heidelberg presses and the Omni-Adast presses use the Presstek direct-to-plate on-press technology and print in a traditional fashion with printing plates. As a result the inks and papers are more conventional and accessible than those for the other digital presses.

For example, the Heidelberg Quickmaster is a waterless SRA3 portrait press capable of handling sheet sizes of up to 13.4×18.1 in. (340×460 mm) at 10,000 sheets/hour. A variety of stocks can be used on the Quickmaster, from 60-gsm paper to 0.12-in. (0.3-mm) boards, including recycled papers, uncoated papers, structured and coated papers, art reproduction papers, and board. In essence it can handle as wide a range of materials as conventional offset presses.

Waterless inks are available from various manufacturers at prices comparable to those of wet offset inks. The makers of dry offset inks include Kast + Ehinger (K+E), Hostmann-Steinber, Toyo, and INX. Special colors can be brought in or produced by combining process colors. An interesting sidenote: Toyo is also the producer of the liquid toner for the Indigo press.

Using dry offset technology permits greater ink densities than wet offset, making it possible to achieve more vibrant, more intense colors. Due to the fact that dry offset involves no dampening solution, which has a cleaning effect, it is advisable to use dust-free papers to prevent hickeys.

# Chapter 12

# Second-Generation
# Variable Printing

In earlier chapters we introduced the advantages of on-demand, digital, and short-run printing. Many of these advantages, such as shorter runs, faster turnarounds, and less-expensive printing are not only possible but deliverable today. On the other hand, there is one advantage that is not readily available or delivered today. That is the incorporation of variable data in color or variable data with pictures for personalization or customization. We call this the next- or second-generation variable printing. In this chapter we focus on the promise of, obstacles in the way of, and current state of the industry of the next generation of variable printing technology.

Let's start with a few definitions and explanations. When we discuss the incorporation of variable color data or pictures into documents, we mean two things: variable printing and color output.

The first criterion requires a technology that permits variable printing. At this point in time that restricts the imaging technologies to either electrophotographic or inkjet. As we discuss in other chapters, the electrophotographic technology is the underlying technology of the high-speed black-and-white copiers/printers (DocuTechs and Lionhearts) as well as most of the digital presses (all except those based on Presstek technology).

The products that can perform second-generation variable printing are those using electrophotographic- or copy-based technology such as the Xeikon-based products (Chromapress, IBM InfoColor 70), the second-generation copiers (Spontane, DocuColor), and the Indigo, a unique hybrid printing technology. Note that we have not included inkjet- or Presstek-based presses (Heidelberg GTO-DI, Quickmaster DI, or Omni-Adast).

We choose not to include inkjet technology within the category "next-generation" variable technology for two reasons. Like the variable data capabilities of the high-speed black-and-white copiers/printers (i.e., DocuTechs), this technology is well worked out and in use, often using data streams that are not PostScript (see IBM chapter on AFP). More importantly, the vast majority of inkjet personalization is one- or two-color printing, not four-color printing, and does not reproduce pictures (halftones) well.

The second criteria is that it is a color technology, therefore eliminating the DocuTechs and Lionhearts. This is not to imply that these devices are incapable of variable printing. Quite the contrary, they are capable of variable printing and are used for several applications. But variable black-and-white printing is a technology that is well worked out, often uses a different printer language (output data stream), and has an established market (corporate and in-house).

## Different Things to Different Markets

"Variable means different things to different markets," says John Sisson, general manager of Banta Digital Services in Needham, Mass. "In the catalog world, I don't know how much sense it makes to create a catalog that is targeted to an individual's area of interest when there are probably only 50 people with the same interest. How variable you need to get is driven by the level of sales you can generate by marketing to a market of one versus a market of 50 or 500."

Banta uses Xeikon digital presses driven by the Barco PrintStreamer RIPs to produce different levels of personalized printing for different markets such as the catalog market, education market, and the general commercial market, which is composed mainly of clients in the health care and retail industries. Banta has developed a custom software package for the catalog market that allows them to create catalogs from a database of products, changing items and prices based on the target market.

According to Sisson, "For us, 'variable' tends to mean every piece is different from the previous one whereas 'custom' means a group of ones that are all the same. For our catalog customers, the value is being able to reach targeted groups of people with different products that they are going to be interested in and not trying to place an 800-page book in front of them for them to pick and choose the items. It's really short-run printing, but to the eventual user, it's a custom or variable product."

"The digital and on-demand presses allow more customization," says Sisson, "but certain applications do not require the additional advantages of the new technology and are better done conventionally using inkjet personalization."

According to Sisson, "For 5,000 individually addressed items, inkjet imaging is still a solution. We try not to participate in things that get into the traditional print arena, because we're not cost-effective and it really doesn't make sense to do it that way."

## Our Definition

Despite the fact that customization and personalization are often used interchangeably, we consider them distinct and separate entities. We define customization as a printed piece targeted at a certain group of people. In contrast, personalization means that it is only going to you.

Here's a few examples:

- You receive a printed piece with your local real estate or life insurance agent's picture on it. That is a customized piece sent to a small group of people based on geography.
- When you receive a document that says "Congratulations you just had a baby, isn't it time you thought about a college fund" that was customized to a small group of new parents.

- When you open your mail and it says "You've turned 60; isn't it time you joined AARP," you received a document that was customized to people in your age group.
- When the professors in the graduate school at the University of Chicago prepare course packs created for each class with specific book chapters, magazine articles, and some notes transcribed by the professor during plane trips, that is a customized package made for that class.
- If Sears recreated it's catalog using a direct response strategy in which the only products in each catalog sent to you where based on your demographics (30-year-old female, married, $35,000 annual salary), that catalog would be customized to that particular demographic.
- Suppose that New York brokerage firm Sanford C. Bernstein creates a portfolio report to its pension-fund clients containing 90% boilerplate pages and 10% specific information that is customized based on your investments. Since it's possible that other people have the same investments, it is not a unique document.

Sometimes the customized content is personally relevant, in which case that document is both personalized and customized. Take any of the examples above and add personal information such as the recipient's name, street address, or a picture and you have created a document that is both customized (going to a small group) and personalized (information only relevant to the recipient).

And other times only one criterion is fulfilled, resulting in a document that is personalized but not customized. When you see your name printed inside of one of your magazines or when you receive one of those documents that say "Congratulations Howie Fenton, you have just won a million dollars," you have received a personalized document (has your name) that is not customized (the content is the same).

## Three Types of Variable Information

You could say that there are three distinct types of variable information: personalized text, customized text, and a customized picture.

Let's use a new car as an example. In this case, the personal information could be personalized text (i.e., your name) or customized with your model car (Ford Escort). Customized text could be a description about the latest model of that car ("The 1997 model . . ."). And the customized picture could be a photograph of the latest model car in the same color or with the same extras (pinstripes) as your existing car.

Based on the data from the direct response experts, the addition of any one of these items — personalized text, customized text, and a customized picture — will increase chances of a sale. Adding more than one of these elements will increase the chances even more.

However, for the service provider, it also adds greater complexity to the database as well as the manufacturing time. In other words, creating a database with the customer name, customized text of the product, and halftone of that car will take longer than a personalized piece with only the customer name.

On the other hand, the more personalized and customized the piece, the greater the sales potential. The obvious question is, "How great is the increase in sales?"

According to direct response advocates, personalization/customization may increase sales from the 2% to 4% for junk mail into double digits. Keep in mind that these direct response campaigns also limit the number of outgoing pieces, which can also decrease costs and increase efficiencies. For example, a 100,000 "shot gun" mailing may be reduced to a 10,000 "focused like a laser beam" campaign.

For example, the promotions that Hilton Hotels Corp. offers senior citizens in its Senior Honors frequent-traveler program features discounts and travel tips. These targeted promotions have persuaded close to half of the club's members to take previously unplanned trips that included stays at Hilton hotels. And Kraft General Foods Inc. says its offers to those listed on its database get much higher response rates

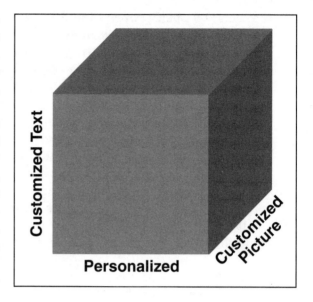

than standard mass-market coupons. The database, says John T. Kuendig, vice-president for market development, is a list of steady consumers who "have a greater value to the brand."

## Range of Complexity

When considering the addition of variable information, it is important to recognize the technology challenges and potential complexities for finishing. On one end of the spectrum are the easy applications, such as short runs with very simple demographic versioning — e.g., one version for men and another for women.

An additional level of complexity would be the addition of names, therefore creating a personalized product. This requires a database of names and some sort of "mail merging" capability. In this case, instead of saying "Congratulations, you've just won a million dollars," it would say "Congratulations Howie Fenton, you've just won a million dollars."

As direct response advocates know, seeing your name in print is a seductive and potent selling tool. This becomes very obvious the first time you see your name printed on the inside pages of a national magazine mailed to your home.

Another level of complexity are the finishing considerations. In this example, the names could be added with or without labels. If labels are used, then a third level of complexity is added: adhering the printed labels to the printed pieces.

Moving further down the line of complexity, we have the multiple versions using customized or personalized pictures. In this case, we could have different text and different pictures selling different products based on some demographics profile. For example, six categories could be 17- to 25-year-old single male, 26- to 45-year-old married male, 46 or over divorced male, 17- to 25-year-old female, 26- to 45-year-old married female, and 46 or over divorced female. For greater personalization, it could also have each person's name.

The ultimate demographic customization and personalization would be to have each person receive a separate piece based on either demographics or previous purchases. In this scenario, everyone gets his or her own unique piece. If we really want to go out on a limb, someday it may be possible to have a picture of each person and see that picture with the product. For example if it is a car, you see what you look like in that car. Or if it is clothing, you see what you look like in that clothing. This is the ultimate example of "one to one" marketing.

## Underlying Motivation

This demonstrates the underlying motivation for using variable information for customized and personalized printing. The interest, attraction, and speculation is used as a direct response tool for distribution within the retail markets. The market estimates are quite impressive. The direct market applications represent a $460 billion market that includes catalogs, TV shopping channels, and other direct-marketing alternatives.

But that is not the whole story. That $460 billion market is only a small percentage (2.8%) of the $2.1-trillion-a-year retail marketplace, which includes supermarkets, mall outlets, car dealerships, department stores, warehouse clubs, boutiques, and much more.

And the direct response portion is growing. According to *Fortune* magazine (April 18, 1994), merchants predict that conventional retailing will remain steady or contract slightly, while the new high-technology marketing such as interactive home shopping channels and personalized direct marketing could achieve 15% of total sales. This would result in revenues of $300 billion — one of the largest businesses in the world.

## Retarded Growth

After reading about the market potential and expected growth, you may wonder what is slowing or retarding the acceptance and delivery of personalized and customized products. Among the factors are the following:
- Engineering issues
- Practical issues
- Implementation issues

## Engineering Issues Affecting Acceptance of Personalization and Customization

A question we are often asked during presentations is why has variable printing succeeded with electronic high-speed printers but not yet with the digital presses. Although there are several clear differences between electronic copiers/printers and the digital presses, the most dramatic may be the output language.

In the last ten years PostScript has become the *de facto* output language or page description language (PDL). As a result it is the printing language used to drive the digital presses.

There is no standard in the high-speed printer community. The manufacturers have developed their own printing languages and translation software that allows the better printers to be used with the more common data streams. Examples of this include IBM's Advanced Function Presentation (AFP) and Xerox's Line Conditioning Data Stream.

There are similarities and differences in the way PDLs and data streams work. In the PDL scheme, the "look" of the document is controlled or "locked in place" by the application (i.e., QuarkXPress). When you go to print the file, the print dialog box is used to create another locked file that is the PostScript version of the application file. This data is sent to the RIP (raster image processor), which converts the PostScript file into another locked file or a bitmap image that is printed bit by bit.

The data stream format workflow is different. There are fewer "locked" file formats and more interpretation at the printer end. This has allowed the data stream format to more efficiently link into databases, extract data, and send this data to the printer, without being encumbered by page layout concerns. As a result the data stream workflows use small files and print fast, but they may not always print exactly as expected.

Two data streams dominate this market: IBM's AFP and Xerox's Metacode. To illustrate our points, we will discuss the IBM AFP data stream. Within the AFP world, the AFPDS (Advanced Function Presentation Data Stream) is the input stream, and IPDS (Intelligent Presentation Data Stream) is the output stream.

AFP is more than a printing output language. It also provides print management capabilities such as archiving, viewing, resource management, and other functions. In other words, unlike PostScript, AFP not only prints files but also tracks files.

Another advantage of data streams over PDLs is error tracking. If you've ever worked as a high-resolution output service provider, you know that the error messages from PostScript files are obscure and difficult to diagnose. For years we tried to guess where the problems were that resulted in "LimitCheck" errors or VM error messages. In contrast, the AFP data stream is designed for data integrity and message debugging.

## Practical Issues Affecting Acceptance of Personalization and Customization

Three practical issues need to be addressed:
- Database
- Cost
- Printing performance

**The database.** The first problems are associated with the database. Is the information in the database? Does it need to be input? Can you get the file off an existing tape? Is it formatted correctly (number of fields, number of characters per field, etc.)? The problem is that, although many companies and printers have databases, they may not be formatted correctly or they may not be compatible with the structure required by the press.

Let's look, for example, at Progressive Impressions, a publisher and printer of customized and personalized marketing publications in Bloomington, Ill. The company recently added an Agfa Chromapress to its existing equipment offerings that include offset presses and high-speed, black-and-white laser printing devices.

A good example of a personalized application at Progressive is a newsletter for a national insurance company. The newsletter is personalized for the local agent, by including a picture of the representative or a map showing the location of the agent's office. This newsletter is typically produced in run lengths in the hundreds. Based on the success of this campaign, the insurance company is considering extending the level of customization with a customer newsletter that targets the needs and interests of the individual recipient of the publication.

According to the president of Progressive, Ken Snyder, "Even though we can't do a fully variable piece now, we can change about 10% of the information on a given piece. But we think it's important to be in the front of variable printing so that we'll be in a good position to offer full four-color variable printing in the future, which we believe is still probably five years away."

The personalization used today is just the tip of the iceberg. For this insurance client, editorial content and photographs could change, depending on the individual consumer's insurance needs. For example, an insurance client with a child who just turned 16 could receive an article about buying car insurance, which could contain a photograph of the consumer's car.

For Progressive to create a newsletter with a picture of the recipient's car, it must first create a database of photographs of every car and, second, it must match the appropriate image to the appropriate client name. Another, often overlooked issue is the additional burden of work created to personalize documents. In other words, creating editorial content such as articles or graphic elements like charts for different demographic groups requires a lot of upfront creative work.

Not only do databases require their own software but for larger systems their own hardware and front-end systems that can store and access that data. Snyder says, "Our largest customer has more than 17,000 agents. To add color photos of all of its agents to the database will require 200 GB of additional hard disk space for storage."

In addition to the hardware and software concerns, managing a database — especially a large database — requires advanced database management skills. These skills are rarely found within most graphic arts companies. As a result, most companies doing variable printing either hire an in-house database expert or subcontract with an outside company to perform the database programming.

And the last issue is, "Who manages and maintains the database?" Because of the confidentially of the information, some companies only provide the information from the database that is necessary to print a particular job, while others may want to outsource the whole data management project. This leads into issues of responsibility, such as who is blamed if something goes wrong. This also makes it difficult to estimate. Are the files prepared correctly? Do we need to check them?

"Whether customers want to do it themselves or not has a lot to do with their level of sophistication," Sisson says. "Outsourcing goes in and out like the tide; some people want to do everything themselves, and then they decide it's not a core business and elect to ship it out."

Similar to other security issues in printing (e.g., pharmaceutical), companies managing information or databases can establish a set of procedures to protect the security of the data. Some companies maintain the information on a stand-alone, non-networked computer. Others provide password protection and/or data encryption. In addition, data can be stored on removable data and stored in a safe.

**The cost.** The second practical issue is cost. As discussed in the chapter on sales and marketing, you don't want to discuss on-demand or digital printing in the same way you discuss traditional printing. If positioned the same way as traditional print, the purchasing decision may be made on cost, and it may lose.

When you consider the costs associated with the database work and the consumable costs, it may not win a side-by-side cost comparison with offset printing. "If customers don't see another value beyond the unit cost, they're not going to be interested," says Banta's Sisson. "Most see a real value in producing these products that make the unit cost less of a factor."

"Sometimes there's sticker shock because the customer's not used to it," says Jim Treleaven, vice president of marketing for Uarco. Uarco has purchased and installed 10 IBM 3170 digital presses in its plants around the country and has ordered another 20. Treleaven adds, "But I don't think cost will be a critical factor in the long run. As more presses are installed, consumables prices will come down, which is a big part of the cost of printing."

In the fall of 1996, Uarco demonstrated — via a live, multipoint video conference beamed to nine of its Document Centers, two public locations, and a site in Europe — its Impressions Showcase '96, a program for managing, producing, and delivering multicolor documents to customers' clients. Uarco's technology partners in this global network of commercial printing and distribution centers include IBM and Xerox.

According to Ranjit Mulgaonkar, vice president of Digital Printing Technology at Moore Corporation's Graphic Research Center, "Very few customers are willing to pay extra for variable printing because they don't know what the return is."

Moore is another multinational company that has installed a number of digital presses such as Indigo and Xeikon in its global operations. "We're working with partners who can help us prove the value of personalization by doing a project and measuring the response rate." says Mulgaonkar. "For this whole area to explode, you need market research that you can take to marketing people to convince them this process has value."

**Printing performance.** The third problem is performance. For nearly a decade the bottleneck in PostScript workflows has been the RIP (raster image processor) associated with imagesetters. Although these problems have been reduced dramatically, there still are performance problems with PostScript devices, and these problems are only related to the printing of static (non-variable) information.

This performance problem will only be exacerbated with the incorporation of variable data. In order of complexity, performance will be affected least by black-and-white text and linework, followed by color text and linework, and lastly by black-and-white and ultimately color pictures.

The issue is that the time required to extract the database information, merge the data with static data, deliver the page to the print engine, and rip the information must be fast enough to keep up with the high-speed print engine.

The state of the art today is the successful incorporation of black-and-white variable information, typically text, in many solutions. Therefore, there are DocuTechs, DocuPrints, and Indigos and Xeikon-based devices incorporating variable fields of black-and-white text.

The successful implementation of continuous-tone pictures is a challenge. Although some manufacturers claim that their systems do not slow down when using variable data with pictures, we have not found any users to confirm that claim. The users we talked to say that even if you store the static portion of the page and give the RIP a head start in ripping the variable continuous-tone data, it still slows down the print engine.

Technologies on the horizon may overcome these performance problems. Adobe has been working on this problem itself with its Supra Architecture and multiprocessing RIPs as well as with others such as IBM with its AFP language.

## Implementation Issues

It is difficult to definitively point to one implementation issue. Because the technology is so new, the only thing we know for sure is that the technology is untested and is in ongoing beta testing.

Although the specific implementation is unique for each device, the theoretical aspect is similar: pages can have static areas that remain the same and areas that are variable, which can be personalized. These variable areas can incorporate text or graphic changes.

Each system has its own way of talking about the underlying elements of customization. These static or unvarying parts of the page are called the "base page," "master page," "fixed data," or the "the form."

Yet for variable printing to become a service that graphic arts firms can offer to a multitude of clients, the industry must first adopt a standard means of transmitting data between the content creator and the service provider. Moore's Mulgaonkar says, "There needs to be a standard way to put a place holder in a document to specify variable objects. There also needs to be a way to link it to the appropriate data in the database."

Although approaches for doing this do exist today, each uses a different method and can be used only with a particular output device. A standardized method would allow various databases, front-end solutions, and output devices to be interchangeable.

"The problem with variable printing stems from the lack of a consistent approach," observes Eric Bean, senior product marketing manager for production systems at Adobe. "Right now, you have to know when you start to lay out your pages who's going to print the thing."

An announcement by Adobe Systems at the Seybold Seminars in Boston in February 1996 may be the first step in standardizing how variable printing is achieved. Along with 26 other vendors, Adobe announced a new printing architecture code-named Supra that is designed to optimize the capabilities of high-end digital printing systems.

Supra is designed to address another barrier to variable printing: the inability of current RIPs to rip static pages and variable data and to feed digital presses at rated speed.

"Today, there's a mismatch between the front end and the back end, and that gap will get bigger and bigger as engine speeds increase," says Mulgaonkar. "Adobe's Supra architecture uses a multiple-RIP system that brings you closer to being able to feed pages to the machine."

## Market Development

Perhaps the biggest challenge for short-run printing in general and for variable printing in particular is the issue of market development. Digital press owners are mounting extensive efforts to educate their customers about the possibilities that digital printing allows and the appropriate applications for the technology.

"To be successful, you have to be willing to develop the market," says Banta's Sisson. "I'd say 86% of the people we talk to about variable printing have the same initial reaction, 'Why are you telling us about this?' However, after brainstorming and getting ideas from them, 68% of those people come up with ideas that are better than the ones we ever had for producing a document that will generate revenues for them."

Selling variable printing in particular often involves a new sales approach. "We have yet to have a program set up through a traditional print buyer," says Progressive's Snyder. "We market to marketers, usually the director of marketing, because print buyers don't understand the marketing value of target marketing."

All of the digital press vendors have identified the need to increase market awareness about the benefits of short-run and personalized printing, and many have completed road shows. Both Agfa and Indigo have performed cross-country road shows.

For example, Indigo extended its roadshow tour of how-to seminars and product demos and offers a 2½-day workshop for sales and marketing professionals at digital printing operations to assist them with pricing and promotion strategies. They also have performed an extensive advertising campaign directed at graphic designers and buyers of printing.

IBM, one of the newest resellers, is planning similar strategies. According to Chris Parker, business line manager for color production printers at IBM's Printing Systems Company, "We believe vendors must take a lead role in developing this market. We view the purchase of variable digital printing systems as a dual ownership, and we don't expect our customers to go it alone."

## A New Architecture for the New Concept

Supra, a new architecture from Adobe Systems for production printing systems, is an extension of Adobe's PostScript language and Portable Document Format (PDF). Adobe developed Supra to address the requirements of the short-run color and print-on-demand markets, including personalization, integrated prepress functionality, and support for postpress components.

Because a PDF is inherently page-independent, Supra will allow documents to be accessed and processed in any order since any given page in the document is independent of all others. PostScript files, by comparison, must be processed in page order. This capability enables page-parallel multiprocessor ripping, allowing Supra-based RIPs to drive digital printing systems at rated speeds and providing the ripping horsepower needed for variable printing. Supra is scalable in that adding more computers or processors to the system increases overall system capability.

The new release of Acrobat, shipped in the fourth quarter of 1996 will add capabilities to the PDF file format required by the high-end color and commercial printing market, including support for PostScript Level 2 color models, overprinting, screens, patterns, fills, and OPI comments. This new version of PDF will serve as the base for Supra products.

Adobe announced that it is working with 26 OEMs to develop products based on this new architecture, including the major digital press and printer manufacturers.

The first products will first focus on the speed and flexibility of multipage ripping. Personalization is not expected to be addressed until 1997.

"Adobe is working with its OEMs on developing mechanisms based on Supra for specifying variable data in front-end creative applications," says Eric Bean, senior product marketing manager for production systems at Adobe. Bean says this will likely take the form of plug-ins to applications like QuarkXPress, Adobe PageMaker, and Adobe Illustrator.

---

## Adobe Supra

Adobe has no personalization software "per se" and has not announced any. It is, however, working on a new print architecture called Supra that will help OEM customers work with variable-data printing. Supra is an extension of the Adobe PostScript language and PDF (portable document format) standards.

Supra is a high-powered version of PDF (portable document format) that allows the rasterization of one file to occur on multiple RIPs in parallel. Therefore it may be possible to send a page with variable information to be rasterized by 4–8 RIPs ,which would drastically reduce ripping times (a feature available on the Windows NT platform from Harlequin).

The architecture is supported by all of the prepress tools needed to prepare digital documents for production: desktop and server solutions for preflighting and color separation, trapping, imposition, and high-resolution picture replacement. Adobe believes that prepress functions should be available for "desktop, server, and RIP" integration to support the broad spectrum of workflows while providing consistent results.

This architecture is applicable to not only digital presses, demand printers, and color printer copiers, but also to large-format imagesetters, proofers, and platemakers. All of these devices share a common requirement: to deliver a large number of monochrome pages or color separations per unit time to the marking engine in the right order. Currently, digital presses tend to produce pages at lower resolutions but higher page rates, while imagesetters operate at higher resolutions but lower page rates.

The architectural objectives include:

- **Scalability.** Adding processing power and system resources to increase throughput.
- **Configurability.** Selecting and including only those pieces needed to meet the system requirements.
- **Multiplatform.** Implementation of the architecture using different computer platforms.
- **Openness.** As interfaces and data structures are solidified and verified for system performance, Adobe will open them up for OEM and third-party development.
- **Quick to market.** Efforts will be made to use existing Adobe and OEM technology to bring products to market sooner.
- **Extensibility.** Evolving the architecture for greater performance and functionality.

Adobe has announced that it is developing a specification for the separation of variable data within its new architecture. This specification will be open and therefore accessible by other manufacturers.

Adobe reports it has achieved page rates of more than 1,000 ppm on a four-RIP system processing jobs typical of the high-end production market. Supra is also applicable to large-format imagesetters, proofers, and platemakers, which require a large number of pages to be delivered to the marking engine at fast rates and in the right order.

## Agfa Chromapress

For years Agfa has been pioneering OPI solutions in its prepress systems, first in the ColorSpace products and later in their OPI server products. For the Chromapress, the database portion runs on Macintosh computers, which limits the database to the File-Maker Pro application. Data from other applications or platforms must be converted into the FileMaker format.

The pages must be created in QuarkXPress with spaces left open for the variable fields as "dummy TIFF" files. Eventually, these items will be temporarily replaced with the variable data using files with the extension "VDF" (for variable data field) and then replaced with PostScript files with the same name having the "PS" extension.

Within the FileMaker application, a set of files is created using the FileMaker Pro report writing features. For each record in the database, a group of files are created for each of the variable fields.

To merge the variable and static information the base page is placed into the printing queue first followed by the variable data. Analogous to an OPI swap, the server-based OPI merge will substitute the variable data for each dummy TIFF file, for each unique page.

The Agfa software, like all the other customized software, is limited by the amount of variable information it can process. There is a limit of 16 variable fields that can contain text or images, per side. Remember that the Chromapress, like all the Xeikon-based products, is a web-based device printing on two sides at the same time.

These fields can contain up to 9 MB of data but cannot exceed 25% of the black portion (plate) or 12% of the four-color area. Variable items have to be approximately 0.2 in. (5 mm) apart form each other and the edge of the page.

This is a common problem with all the Xeikon-based machines, which is a reflection of the Xeikon page buffer that provides two 9-MB areas for variable printing.

## IBM

At the On Demand show on April 23, 1996 in New York City, IBM announced the availability of personalization software called Variable Data System (VDS). VDS allows the IBM 3170 to print variable data in both black and/or full color. The software, which runs on RISC System/6000 servers under the AIX operating system, controls local and remote print submission, job ticketing, and job management. Customers may also choose to install an optional RS/6000-based library server for data on-demand.

According to IBM, printing with VDS does not slow down the IBM 3170. Customers can change up to 50% of the data in full color and 100% of the black on one side or 140% of the black on both sides of every duplex page while the printer is running at 35 pages per minute, 70 ipm.

The IBM variable data creation tools allow document owners to create and implement variable data into their documents (using QuarkXPress and PageMaker). IBM leaves the variable data separate from the master data until the job is ripped.

The Electronic Collator electronically arranges the pages in finished sets before printing, eliminating off-line collation. In contrast, traditional collators are mechanical devices. The collator is capable of storing up to 32 GB of data, equivalent to about 500 pages of a typical magazine. This capacity enables the IBM 3170 to print collated documents with full variable information on A4-size pages at 70 ipm. With text-only documents, the collator can store from 5,000 to 10,000 pages. The Electronic Collator feature also extends variable data to the entire page — 100% of both sides can change from page to page.

MergeDoc is a IBM variable-data production tool that preps and checks variable-data jobs for the RIP. A version of MergeDoc is available for the IBM 3170 with the collator feature. IBM offers a graphical user interface (GUI) for its collator feature, whereas competitor's versions are command-line driven.

## Indigo

Like the Agfa solution, the Indigo personalization is also based on the QuarkXPress application. Indigo offers a Quark XTension that adds a "personalization box" tool to the QuarkXPress tool palette. Another answer to this solution allows users to create pages entirely within QuarkXPress. A maximum of 50 variables can be added per job. According to Indigo there is no upper limit to the number of megabytes of data used for the variable data.

The Indigo software creates a multipage file with one page per database record containing only the variable data. These variable-data-filled pages are then merged with the base page during printing. The base page is rasterized and saved on a hard disk, and only the variable portions are ripped and then merged for output.

## Barco

Barco's PrintStreamer software is available for the Xeikon engines. PrintStreamer is both a hardware and software solution. The hardware includes high-performance RAID (redundant array of inexpensive disks) with a specialized disk controller and interface. The advantage of the PrintStreamer solution is that it addresses the data transfer bottleneck discussed earlier by allowing transfers to occur at the bus speed of 100 MB/sec.

Another advantage of PrintStreamer is that it uses a large storage disk that makes it easy to reprint jobs. This is the same advantage offered by the extended storage option on the DocuTech. Once files are saved in a ripped format, you don't have to worry about pages reflowing. The ripped files can be thought of as final film or plates. Each time the pages are printed from the disk is similar to creating a new original, in contrast to reusing printing plates that wear out or making a copy of a copy that degrades image quality.

Barco personalization software is called VIP (Variable Information Printing). It is based on Barco's VIP Script, which tells the system where on the page the variable elements are stored. According to Barco, VIP scripts can be created using QuarkXPress XTensions or from various databases.

## Scitex

Although shown at several shows over the last year including Drupa, Graph Expo, and Seybold, Scitex announced the "official U.S. launch" of the Spontane at the On-

Demand show in New York City in April 1996. Although officially launched and in beta testing, the product became "officially available" in the fourth quarter of 1996.

They cautioned, however, that even though the official product launch would occur soon and that most of the variable software is almost ready, the variable-data capability would not be included in the "official product release," most likely to occur during one of the 1997 trade shows.

Scitex is positioning the variable software in an interesting way. Scitex says that printing on-demand is not about the speed of the engine; it is about applications. That makes sense considering the Spontane is only a 40-ppm device. The five application areas Scitex will focus on are direct marketing, catalogs, technical publishing, specialty printing, and Internet printing.

Scitex brings two unique features to the development of its variability software: a R&D team experienced in the development of a wide variety of products and the experience of the Scitex Digital Press (SDP) group. The R&D team working on the software is the same team that created the Visionary "link" a decade ago that connected Macintosh computers and the proprietary Scitex system. They also worked on the Catalogic catalog publishing system and the still-unreleased NetDog system for publishing on the web.

More important may be the experience of the SDP staff. The SDP facility, formerly known as the Kodak Inkjet Printing Facility in Dayton, has been working with variable printing for years. Arguably, SDP staff have the most experience in the industry or perhaps the second most experience, with the IBM staff having more experience.

## Xeikon

Unlike the Agfa solution, Xeikon's "Private-I" personalization is not an OPI solution. Instead, it is based on the native dBase application, which is the dominant Wintel (a contraction of "Windows" and "Intel") database format. One advantage to this approach is that any page layout program can be used as long as the page can be saved as an EPS file.

Once the new page is created, it is opened by the Private-I software with two separate layers: the master (static) and variable data. A graphical user interface allows the page creation to occur with a layout tool controlling the variable boxes. These boxes have common names such as <<address>> and <<greeting>>. Once placed, the typographic codes such as font, size, color, and alignment are applied to these fields. After the page layout is created, these fields are linked to database fields using a "drag and drop" technique. The software also has a confirmation process called "check" that confirms that no data exceeds the space allocated to it.

The Xeikon software has the same limitations as the Agfa software: 16 fields per job, 25% black, 12% four-color, 9 MB per side. An alternative solution for the Xeikon is the Barco PrintStreamer.

# Chapter 13

## Accepting Digital Files

Telecommunicating, or accepting customers digital files over phone lines, is one way that service providers can differentiate themselves from the competition. In this chapter, we will discuss the critical importance of fast file transfers for on-demand printing, case histories demonstrating different ways to deliver the service, and the basics involving transferring and accepting files.

In Menlo Park, Calif., A&a Printers and Lithographers measures its telecommunication success using the Internet by its ability to work closer with customers. "We believe it's important to be connected to our customers from the inception of a job to the conclusion," stresses Robert Hu, president and CEO of A&a. "We want to share information seamlessly back and forth between clients and ourselves."

Before the Internet, A&a Printers used a private network to communicate with customers. According to Hu, "This posed an obstacle for many businesses because they couldn't get over the idea of being tied down to a single vendor. With the emergence of the open system of the World Wide Web, it became much more palatable for customers to connect to us electronically."

A&a's customers can enter potential job specs and request an estimate. One of the goals is to develop an instant quote service. On another "page," customers can view A&a's production schedule. Another goal for A&a is to create a more detailed view of that schedule, allowing customers to block their own work schedules. "It will be similar to booking your own airline ticket," says Hu.

"This model calls for a different type of relationship with customers," says Hu. "Being this open is similar to having a closeup picture taken; every flaw in your skin is going to show up. It also takes more forward-thinking customers to be willing to work this way. However, this type of electronic connection is ultimately good for all. It's strategic partnering in the fullest sense of the word.

Another company using innovative telecommunication technology is Graphics Express. Graphics Express, a color prepress shop in Boston, has set up a dedicated T1 line — a powerful 24-channel digital transmission line — capable of transmitting digital files at 1.5 MB per minute. A direct two-way link between the client and Graphics Express's main production facility allows the client to read, access, and select images that Graphics Express has scanned and then stored on a Sun server. Afterwards, the client incorporates the images into a page layout and transmits the file back to Graphics Express for final film output.

Routers control access to data across the line and into the client's and Graphics Express's networks. "Our server shows up to the client as an AppleShare volume," says Rick Dyer, co-owner of the business with Isaac Dyer and Rick Theder. "We'll dedicate a 1-GB drive on our server (for downloads)," he adds. After the files are downloaded to the volume, a hard copy with vital information such as job turnaround time is faxed. "We have one client using the T1 service, and many others using our Switched 56 line," Dyer says.

About the T1 line, Dyer says it does affect Graphics Express's network performance. "You now have extra nodes that the server has to support and route data through. You need a lot more horsepower on your server. Our server is a 25-GB Sun server with on-line storage. We've put in a second motherboard, which means we have a second CPU for backup."

## File Transfer for On-Demand

As we have discussed in previous sections, one of the important, perhaps vital, motivations for on-demand printing is fast, convenient output. Anything that interferes with the delivery of fast, convenient products creates a significant hurdle in the delivery of this product. Therefore, the inability or time delay to get paper (as discussed in the paper chapter), a delay in delivery of the file, or late changes to that file would also create a problem.

Efficient file transfer solutions must be implemented by the printing company using on-demand printing technology. Of primary concern is that the file reaches the printing company intact and that the printing company makes it easy for customers to submit files. In a corporate in-plant printing operation, a company local-area network (LAN) is already in place.

The in-plant printing operation can receive files from its customers in three ways:

- The customer can simply transfer the application file to the in-plant print shop. The print shop must have the same software program originally used to create the job, and the version of the program must be at least as new as the customer's version. A myriad of issues, such as fonts, becomes important, as the in-plant printer will take the customer's file, open the file on one of the print shop's computers, and prepare the file for printing. Working with application files has many disadvantages, but it also has one major advantage: the in-plant printer can make changes to the job.
- The customer can issue a "print" command within the application used to create the job. The LAN fools the computer into thinking that there is a printer attached to the printer port of the customer's computer. In reality, this printer may be located down the hall, in the basement, across town, or across the world. This method works best if the in-plant printer is located within the same building and if the print job is small. There is often no "receipt" that the in-plant printer ever received the job at all.
- The customer can perform a two-step process by first creating a "print to disk" file on his or her computer and then transferring the resulting file over the LAN or via floppy disk (or other removable storage media) to the in-plant printing department. This method also allows the customer to keep a copy of the print-ready file on his or her local hard drive for later resubmission. "Printing to disk"

is commonly referred to as creating a PostScript file, although the print file may be in the PCL format instead of the PostScript language. Professional designers have used this method to submit projects to commercial printers for years. The printing company does not have to be concerned with what application the job was created in or what fonts were used. The printing company simply sends the file to its printing machine for output. The disadvantage is that the printing company cannot correct an improperly created file. The printing company also cannot make editorial corrections to the file.

Using any of the three methods, a customer can take full advantage of a high-speed centrally located printer. Since digital information is being used, the location of the printing machine is not dependent on proximity to the customer anymore. The important criteria now is how close is the printer to the location of the final end user. The ease of transferring digital data transforms printing from a "print and distribute" attitude to a "distribute and then print" ability. Working in a closed corporate in-plant environment has several advantages, the foremost being standardization of items such as paper size, applications accepted, and fonts. Adhering to these standards will facilitate the printing of customers' files.

As more and more company employees work out of a "virtual" office, which may be from their home office or from a notebook computer in a hotel room, the need for a wide-area network (WAN) becomes more important to the in-plant printer. A WAN, an extension of the corporate network, can use standard telephone lines to transfer data or can utilize higher speed methods. Being able to accept files from detached customers is also a concern of the commercial printer, not just for on-demand printing.

The most common method of receiving files from distant customers is via "plain old telephone lines" (POTs). Using a standard modem to transmit files has been a common practice since the early days of computing. The speed of the transfer becomes important if either the modem speed is slow or the file size is large. Black-and-white on-demand printing usually generates relatively small files, normally around 1–10 MB. The current standard for modems is 28.8 baud speed, which translates roughly to 28,800 bits per second. If a 28.8 modem is used, a 10-MB file would take approxi-

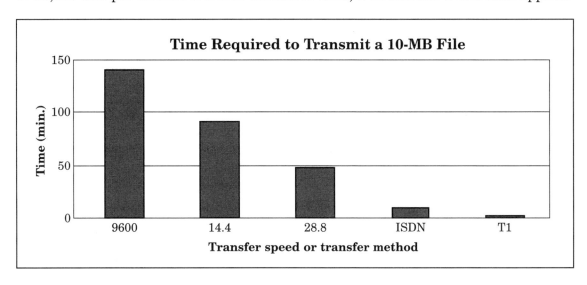

mately 40 min. to transmit. Transmitting files via modem is the most common method in use today because of the relative low cost of the hardware and software.

If faster transfer speeds are required, Integrated Services Digital Network (ISDN) can be used. With ISDN service, the higher bandwidth or speed allows a file to be transferred in less time. Because ISDN is a "dial-up" service just like a conventional modem, it is also well suited for the on-demand industry, because it allows the printer or service provider to accept files from a multitude of different customers at different locations. ISDN is becoming commonplace in the printing industry, especially if the work is in color. Color files can become quite large, sometimes 50–300 MB in size. Because of deadlines, file sizes of this magnitude must reach the printing company quickly. In addition, the cost of transferring these files must be considered. Because of the complexity of installing and maintaining ISDN lines, several companies provide turnkey solutions:

- Digital Art Exchange (DAX) is a company that, in conjunction with AT&T, has developed software and hardware packages to implement ISDN service.
- NetCo WAMNET is a company that will transfer files between printing companies, service bureaus, and advertising agencies, using high-speed data communication lines. A user of WAMNET must be a "member" to use this service to send files to another WAMNET member.
- 4-Sight iSDN Manager is a hardware/solution that provides the necessary hardware to connect to an ISDN line. The package also contains software that enables billing information and remote soft proofing.
- Adobe Virtual Network has assembled a suite of software products that facilitate the transfer of files over not only ISDN lines but also over phone lines and the Internet.

Several organizations have been formed that will facilitate the file transfer and printing of jobs. Worldwide Electronic Printers Network (WEPN) and International Printers Network (IPN) are two groups of printers that have made the process of having remote printing accomplished easy. A customer works with a local member of either organization and can arrange to have the printing done around the world at remote locations. The savings in not having to ship printed paper can be enormous. For overseas printing, the usual delays in customs processing are avoided, saving a great deal of time.

Printers who routinely receive a large number of files or large files from one customer may choose to install dedicated transmission lines. These high-speed, expensive lines are referred to as T1 or T3 lines. They offer extremely fast transfer times but are dedicated to a particular customer site. When working with large color files, T1 or T3 lines are the fastest possible methods.

When receiving files from a customer for the on-demand print industry, a simple method must be implemented to ensure that the file is received and that the customer's production requirements are understood. An electronic job ticket is highly desirable.

The simplest form of this is an electronic message that travels along with the print job. An electronic Bulletin Board System (BBS) is a common method that a printing company will implement to allow a customer to send a file and attach a message to it. A BBS also allows a customer to retrieve files from the printing company.

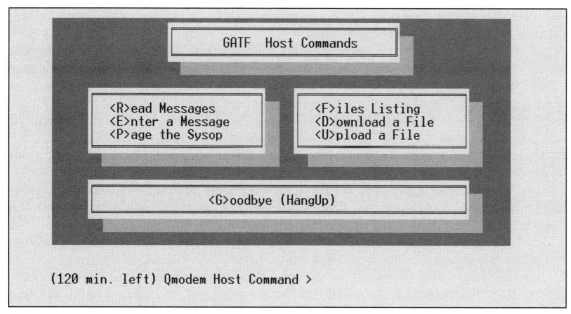

Menu from a Bulletin Board Service (BBS).

All a customer needs is generic communication software, a modem, and a telephone line. If a customer needs a particular print driver or a particular PostScript Printer Description (PPD), it may be available directly from the printing company via its BBS. The advantage to the printing company is that setting up a BBS is relatively easy and inexpensive.

A dedicated phone line is desirable so that the customer can access the BBS anytime of the day. However if the customer is making a toll call to access the BBS, long-distance charges will apply during the file transfer.

Another method for a printing company to accept files with a text message is to subscribe to a commercial on-line service such as America Online or CompuServe. Subscribing to either of these services will allow the customer to send files to the printing company's screen name or number and attach a message to the file. If the customer sends the file via an Internet account, the printing company will still receive the file. These commercial services provide access to the Internet. The advantage to these services is that it is almost always a local telephone call.

A printing company can subscribe to an Internet service provider (ISP) to simply have electronic mail (email) capabilities. This will allow them to receive files from a broad range of customers. Printing companies, especially on-demand printers, have started to have their own web pages on the World Wide Web (WWW). These pages not only serve as advertisements for the printing company but also allow customers to submit files easily. Some of the more outstanding pages also allow a customer to have a price for a printing job quoted.

Xerox makes the InterDoc system to facilitate accepting files over the WWW. Through a printing company's home page, a customer can submit files, fill out an electronic work order, check on the status of previously submitted work, and order reprints. The system runs on a Sun Sparc workstation and includes Oracle database software. An administration software program allows the printing company to maintain information about customers' work.

Making the process of transferring files to the printing company easy is the key to a successful digital printing operation. Several companies provide software that makes it extremely easy for the customer to fill out an electronic work order, gather the files to be sent, and initiate the file transfer over a LAN, WAN, modem, or Internet FTP. PagePath Technologies LAUNCH! and ASAP! from Microbeam Corporation are two products that are being used extensively in the on-demand industry.

These programs consist of two parts: (1) a receiver program that resides on a computer at the service provider and (2) a sender program that is distributed to customers. The receiver component is used on a Macintosh or Windows computer and is used to receive the customer file. Once a file is received, it can be decompressed and then sent into the production cycle. The file that is transmitted contains a text file which is typically printed out and used as a work order or job ticket. The receiving computer is also used to customize and manufacture the sending software.

This sending software, which is tailored to the customer, is usually supplied on a 3.5-in. (89-mm) floppy disk for installation on the customer's computer. Specifications, such as unique papers or bindery capabilities that the printing company offers, can be built into the sending software. This type of "personalization" can also act as a form of advertising for the printing company, showcasing their particular capabilities.

Once the sending software is installed, the customer simply gathers the files to be sent, fills out an on-screen job ticket, and the software will automatically compress the file and transfer the job to the printing company. The job ticket information can be saved and later reused for other jobs being sent.

ASAP! from Microbeam Corporation.

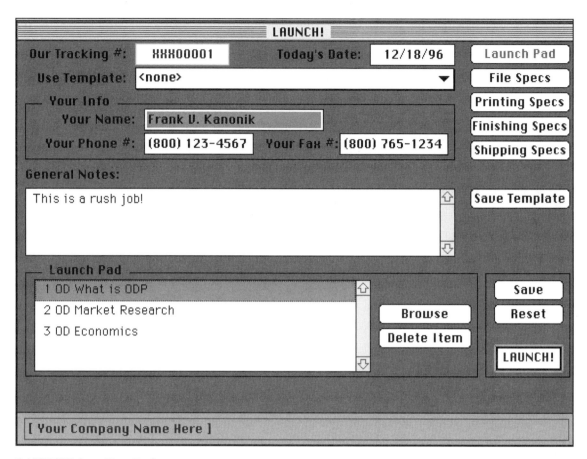

LAUNCH! from PagePath.

# Chapter 14

# Networking

Networking is becoming a critical factor in creating and delivering documents internally and externally in both printing and publishing organizations. There are several definitions of networks. Most of the time we think about networks as internal equipment and services. As a result when people discuss networks, they describe problems with LANs or local-area networks.

But there are networks that connect different sites or WANs (wide-area networks) as well. Many of the larger printers already offer high-speed data transmission between tier sites. Some of these services were described in the section about printers offering print on-demand services. In this chapter, after discussing LANs, we discuss a service from AT&T that could be used by smaller companies.

Basically there are three types of LAN networks used with demand equipment: one that basically serves an output service, another that serves a publishing service, and one that serves an in-house reprographics department. Some are used more for output, others are used more for publishing applications, and some perform both tasks.

## Networks for Printing

College campuses with on-demand equipment use the networks almost exclusively for output. College campuses using on-demand printing create and manage millions of impressions each semester. For example, Stanford University Bookstore, one of the largest campus custom publishing operations in the country, regularly coordinates over 6 million impressions for 13,000 students.

Managing and saving these files is a Herculean task. At Stanford, Dan Archer is developing a campus portfolio of pages from which to draw material for course packs. At Stanford the campus bookstore is networked with the campus printing facility as well as with three off-campus vendors. When the in-house capabilities are added to the out-of-house capability, the campus store has network access to seven DocuTechs. That is the only way they can create all the course packs required during a six-week window for the 13,000 students.

Considering that these course packs are basically books stored as a large, bit-mapped file format, these files get incredibly huge. What many people don't understand is that designing, configuring, installing, and maintaining a network for printing and publishing is entirely different from working with a local-area network in an office environment. The needs and requirements are much different. Probably the biggest factor that distinguishes a printing network is the size of the files.

## Networks for Publishing

Publishing networks are not the same as networks used just for printing. One of the most important considerations for a publishing network is the pressure of deadlines. While setting and meeting deadlines are the responsibilities of the editor and production managers, these pressures also mean that a publishing network must be healthy at all times; or, at the very least, the publication should still be able to get out the door even if the network crashes completely.

Publishing networks are also different in that a variety of creative people — writers, editors, graphic artists, art directors — are all working on the same project, all working toward a common goal. The network must make information available to these creative people, and they must be able to pass data back and forth with a minimum of effort. In other words, the operation of the network should be as transparent as possible.

Working with networks is a moving target. New software and hardware is introduced daily. Your basic concerns remain the same: configuration, installation, and maintenance all follow a set of basic rules that you must keep in mind.

Personal computers have radically altered that scenario with what has been described as the "fourth wave." The first wave was hot-metal, second phototype, and third proprietary integrated systems. The fourth wave consists of:

- Mainstream computer hardware
- Mainstream operating systems
- Mass-market application programs
- Computer industry standards for interfacing devices, networking, and exchanging data

Such off-the-shelf components, based on open architecture, provide almost unlimited flexibility. This much choice, however, creates its own set of headaches. As network manager, you will be expected to recommend specific hardware and software products to buy from a bewildering range of choices. Then you'll be expected to ensure that it will all work together.

## Networks for In-House

Desktop automation should give employees access to information stored in documents and databases throughout an organization. Networked document imaging systems should provide access to microfilm, optical disk systems, CD-ROM discs, or mainframes.

These systems need to deliver documents to desktop workstations so that workers can massage, and augment, that information in popular desktop software programs. The imaging system should also support a variety of output devices, so the user can select the one that's most appropriate for each document.

The ability to share information is a powerful tool. A networked document imaging system equips employees, for the first time, to take full advantage of existing information. It improves the productivity of not only individual employees, but the entire enterprise.

No matter what a company sells — whether it's a product or a service — the quality of information is a critical factor in its success. Networked document imaging software allows desktop users to capture, store, manage, and share information among all networked devices.

The foundation for any document imaging system is a network architecture that links devices through support of open standards and industry standard communications protocols. Modular application software is the next step.

With these two building blocks in place, each company can customize a document imaging system to meet its needs for the present and in the future. A modular design allows for upgradeability as needs dictate and budgets permit. Adherence to standards protects the current equipment investment.

There are many types of document imaging software. To achieve widespread access to information and swift delivery of documentation, software should provide certain benefits:

- **Accept and serve images from a variety of formats.** The software needs to serve documents to employees in popular page description languages like Post-Script and PCL. These computer-generated documents should be stored and delivered to employees in the original page description language to prevent the need for reformatting. Imaging software also needs to provide access to images in TIFF and CCITT Group IV formats.

- **Scan existing paper documents.** Most companies have warehouses of paper files that need conversion to microfilm and electronic formats. In the past, companies have not been eager to scan documents because imaging systems provided such limited access. Flexible imaging software delivers images to users' desktops in seconds. Paper documents can be scanned to the system and stored on optical disk jukeboxes or other electronic formats and archived onto microfilm. Microfilm images can be digitized and stored electronically on the network as well.

- **Support all standard desktop platforms, including Sun, H-P, Digital, PC-compatibles, and Macintosh.** Companies can install software that permits desktop users to transmit document requests electronically. Some documents may be printed, then routed to employees. In other cases, documents may be posted to centralized electronic storage devices for immediate retrieval. Systems that boost productivity must provide desktop users of all platforms with swift access to images.

- **Merge images with text.** Networked document imaging software should allow users to download images and combine them with graphic and text files to create compound documents. Typical applications include marketing proposals, presentations, client newsletters, and product sales sheets.

- **Perform basic editing functions.** In most cases, minor cleanup and editing is required. Notes in the margin need to be erased, as do holes created when three-hole-punched paper is scanned. Graphic images need to be reduced, enlarged, cropped, and rotated. Adjustments need to be made to resolution, intensity, and contrast of halftone and continuous-tone color images. Desktop publishing programs provide advanced editing functions, but networking software should make basic editing functions available to all users.

- **Manage the electronic library.** Companies must preserve the integrity of documents. Networking software must support library management functions like indexing and controls. When employees request and retrieve a document, they must be given a copy, not the original. Typically, only a system administrator can check new documents into the electronic library. Library management software may also create title pages, table of contents, and page numbers for documents when they are stored.
- **Output documents to a variety of networked devices.** When employees have created a document, they need to select the appropriate delivery format with the receivers of the information at the forefront of their minds. Fully featured networking software should permit documents to be delivered to user workstations, central storage systems, or networked printers. It should also support output to microfilm, microfiche, Photo CD discs, and writable CD discs to permit archival records and easy distribution to users.
- **Manage on-demand printing services.** Employee directories and manuals of all types require rapid on-demand printing. Sales handouts, proposals, and other documents require on-demand printing in much lower volumes. Networked imaging software should allow employees to input finishing and delivery instructions on an electronic job ticket, and specify the type of printing required. Large-volume orders can be routed to high-speed printers, while smaller orders are sent to mid-volume units.

In the past, most users have only had access to low-volume laser printers on their local-area network. If multiple sets were needed, they often printed a master, then walked to a copier to create duplicates. On-demand printing allows employees to output finished sets without leaving their office and still enjoy the benefits of original output quality.

Networked document imaging systems should give users access to accurate, timely information so they can efficiently create new documents of value to the organization. Why go to the library if you can only check out one kind of book? A better question is this: Why implement a document imaging system if it does not provide desktop access to files and electronic databases throughout an organization?

## Ideal Network

As an exercise to get started in the right direction, let's look at the attributes of an ideal network:

- The network should work. Normal operations by users should not cause the network to crash.
- The network should be simple to configure, connect, and use. A minimal amount of technical expertise should be enough for a user to take advantage of the network's full range of capabilities.
- The network must be flexible to adapt to the needs of a growing organization. Adding new devices, moving, and rearranging the network should cause minimal disruption. Open networking architectures allow for expansion and alteration.
- Speed and bandwidth must be adequate to prevent delays and bottlenecks. Otherwise, the whole purpose of the network is lost. This becomes particularly

acute with a publishing network since, as deadlines approach, more and more people are using the network and demanding high performance as they attempt to complete the publication.

- Security demands that the network be safe from damage. A simple, yet secure method of controlling network access should be available. The network should allow different levels of user access. Users need private as well as public files, and certain information must be protected. In addition, a network must provide an easy means of protecting existing information storage with backup and copy routines.
- Low cost per connection and for maintenance.
- Reliability and data protection ensures that when the network fails, and it will, there are procedures for a "soft" crash allowing users to save their work before total failure.
- Interface support should support a method by which dissimilar networks can be interconnected.
- Broad support services should include methods for information transmission and sharing of printers and other peripheral devices.

On a more specific level, it's vital that your network will:
- Support the number of nodes required.
- Cover the physical area of the site.
- Cope with expected traffic.
- Meet standards of compatibility for operating systems and network communications. Different equipment should be able to communicate and interact through the network.

## WAN: Wide-Area Networks
A wide-area network is a network typically extending a LAN (local-area network) outside the building, over telephone lines, to link to other LANs in remote buildings in possibly remote cities. It can operate on POTS (plain old telephone service) lines or leased lines from one analog phone line to T1 (1.544 Mbps).

At the On-demand show on June 27, 1995, six leading prepress, printing, and software companies announced plans to work with AT&T toward the goal of making their printing and software products compatible with the recently announced AT&T Network Demand Printing (NDP) Service.

AT&T signed agreements with Agfa-Gevaert Group, Eastman Kodak Company, Indigo N.V., and Scitex Corp. Ltd., and a letter of intent with Adobe Systems, Inc., and Quark, Inc. The companies plan to evaluate and adapt their print technologies to operate with AT&T's planned service.

Today the service connects with Xerox DocuTech Network Printers. Customers will be able to use the service to distribute and print documents, such as books, manuals, brochures, price lists, and marketing materials.

Earlier in June 1995 AT&T and Xerox Corporation announced a market trial of a network-based printing service, which speeds the distribution and printing process for publishers, printers, and businesses that do high-volume printing to multiple locations. The service uses the AT&T Network to send documents to be printed at customer-specified locations on compatible printing equipment. AT&T's intelligent network monitors and manages the distribution process.

According to Jim Cosgrove, vice president and general manager of AT&T's Business Multimedia Services, "We plan to offer an anytime, anywhere, on-demand printing service. We envision the day when, with the touch of a button, a user simultaneously can send high-quality documents across our network to many locations to be printed on 'user-specified' equipment from many different vendors."

In addition, customers will not be required to hire specialized technical experts to plan, implement, and administer complex networks. AT&T will take care of capacity planning, network and software upgrades, and assume full responsibility.

## Flexibility
During the initial market trial period the service supports black-and-white documents. By 1996, AT&T expects to provide color capabilities, online document storage, and directory services — the first of which will list available destination sites.

Besides the printing, customers can specify finishing requirements, such as the type of paper and binding as well as destination sites. The network distributes each job and automatically sends a status report back to the originator when all jobs have been completed.

According to Cosgrove, "Businesses today face shorter-time-to-market pressures, cost-reduction issues, and a critical need to rapidly access and distribute documents. Premises-based networks met past needs, but globalization, decentralization, and electronic bonding with customers, partners, and suppliers are forcing new communications paradigms."

# Chapter 15

# Bindery Issues

In an attempt to increase productivity in any form of manufacturing, the first step is to identify your bottlenecks and develop strategies that overcome or work around the bottlenecks. However, after you address your bottleneck, your overall throughput may only increase moderately because another previously masked bottleneck appears. You may be wondering what this has to do with print production and the bindery? The answer is that the increased prepress and printing capacity, resulting from technologies such as digital printing, direct-to-plate technologies, and on-demand printing is shifting the bottlenecks away from the actual printing process to other places such as file transfer, order entry, customer service, and the bindery.

For years we have attempted to eliminate the bottlenecks in print production by focusing on the following areas:
- Manual stripping
- Electronic retouching, trapping, and imposition
- PostScript RIPs
- Press speeds

Now that these printing bottlenecks have been reduced and makereadies have been accelerated due to new technologies, the location of the bottleneck has moved. One of the new bottlenecks is in the binding and finishing of all these pages. In other words, the bottleneck in print production has moved from prepress and printing to the bindery.

Another motivating factor underlying the interest in the bindery is a renewed concern about employee safety and health, which has prompted new alternatives to replace lifting, cutting, and sorting. The tradeoff: employees could run the machines via keyboard rather than by hand.

And lastly, there is the realization that the bindery may be the most important part of the printing process. Why? Because if the pages are cut or bound wrong, it must be reworked from the beginning at very high cost. As a result of these motivations, new solutions are becoming available and a renewed interest is being focused on the bindery.

## Bindery Upgrades

For over two years, Fort Dearborn Litho (Niles, Ill.) has been updating its bindery department. According to Nick Adler, vice president of operations, "We've been replacing our entire bindery. Our equipment was older, and we wanted to provide a safer, more ergonomic environment for our employees. We're adding more automated equipment to improve efficiency and quality, and to try to reduce handling, both before and after the cutting step."

As described in the label chapter, the label industry is experiencing growth and may become a prime candidate for on-demand or digital printing technology. Fort Dearborn, as well as its sister companies, produces labels for the packaging industry, mainly for food and beverage products. The other plants include Eagle Printing, a 100-employee plant in Michigan; Graphic Arts, a 165-employee shop in Texas; Flextech, a flexible packaging plant; and Virtual Color, an 80-employee digital prepress and design firm in Itasca, Ill. Fort Dearborn's labels appear on retail and grocery store shelves or are used by packers.

Like other companies with high-volume bindery concerns, Fort Dearborn sought a complete equipment solution. They installed off-line joggers for sheets, which can then feed work to air-table conveyors feeding the cutters. In addition, automatic lifters were installed to deliver jobs to new Polar Type ED guillotine cutters, which were added to improve productivity and makeready and are equipped with an Autotrim on-line waste removal system. Lastly after the job is cut and trimmed, an automatic banding system secures the product until final packaging.

## The Need for Speed

According to Adler, "Our bindery equipment was generally about 30 years old. Speed was an important consideration, and we've gained in efficiency. We've been able to increase our output with half the resources."

"We realized we were definitely overlooking the bindery. We print and cut labels, so the pressroom and bindery have everything to do with the function of our output. Unfortunately, in the past, our bindery was set up to do everything but be efficient, but by standardizing in the bindery, we've increased efficiency."

Perfect Binding (Indianapolis, Ind.) was one of the first binderies in the U.S. to be equipped with a Kolbus Systembinder 2000, a high-speed adhesive binding line. Perfect Binding, a 16-year-old, 55-employee company, is a leading binder of annual reports and books. The new machine, rated at 10,000 books per hour, completes the equipment tool kit that also includes mechanical binding, wire binding, and saddle stitching.

According to says Jeff Combs, the vice president, "Within two weeks, we had the Systembinder 2000 up and running at full capacity on a two-shift basis. It has reduced our makeready dramatically. Standard setup time on conventional equipment is 90 minutes, but we can set up the Systembinder in 5 minutes. It is equipped with four Hewlett-Packard computers and a color graphic screen that shows job status or stop reason. It's so easy to use — all we do is load covers and go — that we should be able to double or triple our job output."

Business is so good at Perfect Binding that 23,000 sq.ft. was added to its 60,000-sq.ft plant for increased capacity, and the company purchased a new tractor trailer to deliver jobs. Combs adds, "I can't believe how much new automation I've seen in one year. Technology has remade the folders, and trimmers can now be set up in three minutes. It's just endless what new automation will be seen in the future. I can foresee touch screens being added to bindery equipment. It could probably be done now but it's too costly, but in five years we'll see it on all equipment."

Another successful bindery using new technology is Bindagraphics, a 22-year-old trade shop in Baltimore. This 121,000-sq.ft. facility employs 210 people and offers a host of varied services including binding, inkjet imaging, gluing, and folding options; loose-leaf products; and such specialty services as tabbed dividers, UV coating, film laminating, foil stamping, embossing, and diecutting.

According to Marty Anson, company president, "Lately we've been emphasizing our inkjet personalization services. We have a new Videojet system on our folder for direct-mail products, a Scitex system on our saddle-stitching line for inside/outside addressing, and a Videojet unit on our polybagging line to add messages or addresses on polybag wrappers."

In terms of future products, Anson is keeping his eye on the new signature recognition systems being developed by Kolbus and Muller Martini. "These systems can save a bindery a lot of problems," he says. "The system tells the operator when the wrong signature has been placed in a pocket."

The importance of databases and the ability of database information to function across computer platforms was described in the chapter on second-generation variable printing. This data may also need to be translated to bindery systems.

One of the larger printers providing database and direct-mail services is Quebecor Printing. "We are striving toward a networked bindery," says Paul Gaboury, director of information systems for Quebecor Printing USA. "We want to have all bindery equipment components connected to each other as well as to a central file server. We're hoping that bindery vendors will support open systems in which bindery machines can gain direct access to each other. We expect to attain this goal and be up and running by the second quarter of this year. We're the test site for the rest of the Quebecor plants."

According to Gaboury, the company has two goals. The first goal is to capture the manufacturing data files used to print direct mail and then store them in a database where they can be analyzed. This information could be used to track what jobs were running and when, and locate the reasons why equipment failed in an effort to better manage the plant and control waste.

The second more ambitious goal is to automate the distribution of the direct-mail pieces. Instead of using the names and addresses supplied by computer disk and then lost, Quebecor wants to access the data from a satellite feed, capture the data, and print the jobs on demand.

"We are seeking to create an electronic cnvironment. Currently, all the data files are lost after the direct-mail piece is printed. We want to save, capture, and analyze the files for statistical purposes to improve our efficiency," says Gaboury.

## Binding 101
The finishing of digitally printed products can be divided into two categories:
- **In-line finishing,** which can be considered automated since the finishing operation is either incorporated into the printing machine or can be conveniently attached to the print machine.
- **Off-line finishing,** which can utilize existing finishing equipment, is considered a separate operation from the printing of the product. In this case, the digital printing machine is used simply as a printing press.

The decision to use in-line or off-line finishing is based on the type of work to be produced, the present finishing capabilities of the print shop, and the possible need for high security, where the printing must remain within the physical confines of the digital machine.

Finishing a digitally produced piece can be as simple as a staple in the upper left-hand corner or as complex as an elaborately glued and folded mailer containing separate pages with tear-out perforations. Proper planning for the finishing operation, while important, is not as intricate as in the traditional lithographic industry. The main difference is the sheet size.

While a digital printing machine can usually image two 8.5×11-in. (216×279-mm) pages on an 11×17-in. (279×432-mm) sheet of paper, a lithographic press may be printing a 23×35-in. (309×889-mm) sheet of paper that contains eight 8.5×11-in. pages. The smaller sheet size, with its smaller number of folds, is easier to work with than the many folds of the larger sheet size.

The traditional lithographic press finishing operation usually produces large quantities of the same image. The on-demand digital printing machine has the capability of producing a different image on each sheet of paper. This leads to finishing operations not possible with conventional lithographic equipment. For example, each page can be personalized with different information such as a name and address for mailing purposes, or the pages of a catalog can be customized with products directed to the recipient. This type of printing poses a challenge to the finishing operation since there can be no rejects due to bindery problems. Careful attention must be paid to ensure that each piece, because it is unique, is properly finished and accounted for. Solutions are available that electronically scan pages as they are printed to verify that the page is indeed printed and in the correct sequence.

The simplest type of in-line binding usually incorporated into the digital printing machine is stapling. The staples are normally fed off of a large roll of wire, ensuring a large supply. The upper left-hand corner is the traditional placement of the staple. A variation of this is called side-stitching, where two or three staples are placed along the left-hand edge of a book. Side stitching gives the publication a more finished book-like appearance; however, the book will not lay flat when side stitching is used.

Another simple form of finishing is three-hole drilling for placement into binders. This is usually performed off-line using a wide variety of traditional drilling equipment. A variation of this is to use pre-drilled paper, which usually adds 1¢ to 2¢ to the price of the page.

Comb binding, which uses a plastic piece to hold the pages together, is used frequently. This type of binding has the advantage of being efficient for very short runs. The process of cutting the holes into the page and inserting the plastic comb through the book can be performed on very inexpensive tabletop machinery to produce a few books, usually in an office environment. For larger quantities, the process can be automated, with larger machines that can drill multiple copies. The plastic combs range in size from 3⁄16 in. to 2 in. (5 mm to 51 mm), which allow the binding of up to 500 pages.

The combs are available in a range of colors to lend a more designed look to the piece. The advantages of comb binding are its simplicity, the option of incorporating a thicker, possibly litho-produced cover, the ability for the open book to lay flat, and the opportunity for the book to be reopened to remove or add pages. The drilling process can be integrated in-line into several digital printing devices.

A type of binding that is similar to comb binding is wire coil binding. The book must be punched in a similar way to comb binding, then a wire coil is threaded through the perforations. This process is used when a more durable bind is required.

Mechanical binding devices: plastic comb bound *(top)* and double wire loop bound *(bottom)*.

Tape binding is a form of perfect binding that can be performed either in-line or off-line. A strip of flexible cloth tape that contains a heat-activated glue is applied to the edge of a stack of paper. The glue will dry or cure almost instantly as it cools, making this process ideal for the on-demand print industry. The tapes are available in many different colors. The Xerox 5090 and DocuTech series of printers incorporate tape binding in-line into their machines and can accommodate page counts from 15–125.

Several manufacturers make off-line machines that apply the cloth tape. The machines can handle book thicknesses up to 1.5 in. (38 mm), and some also allow the contents of a book to be changed by reheating the tape and swapping out pages. The cloth tape type of finishing is durable, giving a very high pull strength. Heavyweight covers, either produced digitally or on litho equipment, can be incorporated into a tape-bound book. Tape binding also has the ability to allow a book to open flat.

Saddle stitching is being performed on both in-line and off-line conventional equipment. With in-line saddle stitching, the pages are printed in the correct order until the book is totally printed. The collated pages are then stitched in the center with two pieces of wire, folded in half, and trimmed on the face edge. The entire process can be automated, with finished books coming off the end of the machine. The typical maximum page count is 88.

Depending on the thickness of the paper, finished trim size can range from 5.5× 8.5 in. to 8.5×11 in. (140×216 mm to 216×279 mm). Some users choose to allow the digital printer to act as a printing press and then perform the saddle stitching on traditional off-line machinery.

In-line saddle stitching is growing in popularity. Since this is true in-line finishing, not only can each book have personalized information but the entire process can be automated. This is important for short run, fast turnaround type of work. Eastman Kodak and Xerox both offer in-line saddle stitching capabilities. If preprinted covers are needed, a cover insertion module available for the DocuTech will insert a preprinted cover automatically to the pile of interior sheets before the stitching process occurs.

ChannelBinding, available from ChannelBind Corp., is a type of binding that mimics traditional hard-cover case-bound books. The interior pages are printed and inserted off-line into a one-piece hard cover. The cover contains a metal channel that, when compressed, grips the interior pages and holds them to the cover. A special tabletop machine performs the compression of the metal channel. The same manually operated machine can also uncompress the metal channel to allow pages to be added or removed.

Covers are available in different spine thicknesses to accommodate booklets ranging in thickness from 0.20 in. to 1.28 in. (5–33 mm). Covers can be off-the-shelf and can be screen-printed or foil-embossed. They can also be custom made by Channel-Bind with a digital- or litho-produced image formed into a hard cover.

True perfect binding, which is a one-piece cover wrapped around a pile of collated sheets, is available both in-line and off-line for the on-demand industry. With perfect binding, book thicknesses can be up to 350 pages. When combined with a digital printing machine, an in-line binder will streamline the production of perfect-bound books. The C.P. Bourg company offers the BB2005 perfect binder, which is used in-line with the Xerox DocuTech. When a complete interior is finished, it is automatically fed into the binder, the spine is ground with a cutting wheel, hot-melt adhesive is applied, and the cover is wrapped around. While the actual binding is in progress, the DocuTech is printing the next interior, ensuring continuous operation. The books that come off the binder are then manually transferred to a guillotine trimmer that will trim the head, face, and foot of the book. The typical sizes can range from 7×10 in. to 9×13 in. (178×254 mm to 229×330 mm).

A prime consideration when exploring the various finishing operations is the paper handling capabilities of the digital print machine itself. The number of different paper sources that a machine can utilize can be the difference between a fully automated system versus one that requires some manual assembly. Multiple paper drawers can contain different color stocks, or they can contain a front and back cover in addition to the white paper for the interior. This is a principal factor when producing manuals that will be three-hole-drilled. In addition to the white paper, there may be tabs, color sheets, or preprinted sheets. Having enough paper bins to accommodate these needs will allow a job to be fully automated without any manual collation needed.

ChannelBind™ System 20.
*Courtesy Boone Business Products, Inc.*

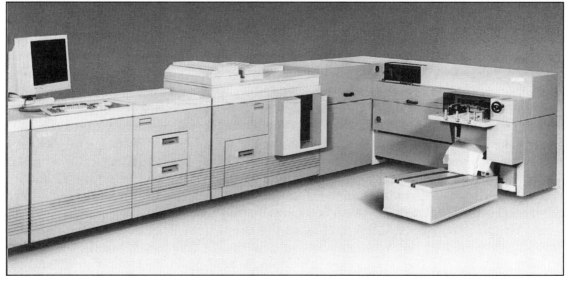

Bourg BB2005 perfect binder *(far right)* attached to a Xerox DocuTech.
*Courtesy C.P. Bourg, Inc.*

Another concern with paper is how much the machine may hold. Having to stop periodically to refill paper supplies can slow productivity. An ample supply of paper, along with the ability of the machine to intelligently warn an operator when supplies are low is desirable. A unique solution to paper supplies is available from Roll Systems Corporation, whose DocuSheeter allows a DocuTech or DocuPrint to operate off of a roll of paper. The 17-in.-wide (432-mm-wide) roll of paper is cut into sheets as it enters the printer, allowing for 12 hours of continuous operation at rated speed.

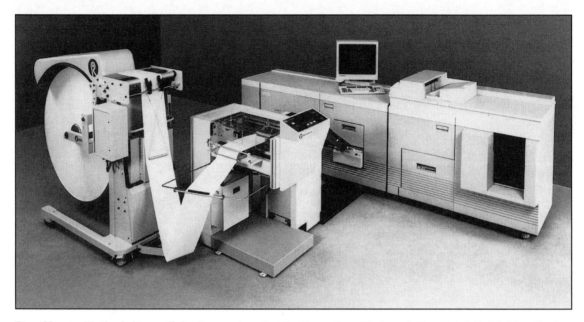

DocuSheeter™ *(far left),* a roll-feeding system for the Xerox DocuTech and DocuPrint Publishing Series. *Courtesy Roll Systems, Inc.*

## New Products

Across the board, at all of the major printing industry trade shows in 1995 and 1996, from Dusseldorf, Germany, to Chicago, manufacturers are focusing on increased automation and productivity.

Heidelberg USA introduced digitally controlled folders from Stahl. According to Heidelberg, the Stahl TD Series is capable of processing up to 8,260 linear inches per minute (210 m/min.), making them the fastest buckle-plate folders in the industry. Also introduced was a folding system that produces thread-sealed booklets. The TD folders are digitally controlled, which allows all major feeder, folder, and delivery functions to be monitored and controlled from a central command post. It also has a total amount batch counter. All of the stations on the TD folders, including the feeder and delivery, run with maintenance-free, frequency-controlled inverters. This allows fine speed control between 2,000–8,260 in./min. (50–210 m/min).

At Graph Expo '95, Heidelberg showed the Stahl USA B-20 as the front-end of a system that produces thread-sealed booklets in line. In this demo, the B-20 was networked with Stahl's FS 100 accessory system, a VSF-52 knife folding unit, and an SKP-66 pressing unit.

The FS 100 uses a special plastic-coated thread to bind folded signatures into booklets. The system also contains a melting unit, which seals the threads on the back of the signature for better appearance and longer durability.

In 1995 the Muller Martini Corporation introduced AMRYS, an automatic, computerized makeready system for the mid-range Prima saddle stitcher. AMRYS is designed to help reduce makeready time. The manual adjustments that are normally associated with saddle stitcher makeready, including size setting and timing, are motorized. The adjustments are made automatically, after entry by diskette from a PC controller or by keypad. All of the feeders on the stitching line can be adjusted simultaneously or set individually. The job parameters can be stored on disk for repeat runs, which gives the AMRYS system unlimited storage capacity.

Many specific parameters can be entered on the keyboard, such as feeder timing to gathering chain; feeder adjustments for length and width of signatures, thickness, side and front stops, signature opening, and signature thickness; stitcher transfer; trimmer infeed, product length and width, head and foot trim, and product transfer; and stacker bin. This feature is of particular value to binderies with frequent job changes; for large numbers of feeders, which can now be set simultaneously on command; or production requirements with many split runs and repeat jobs. AMRYS cannot be retrofitted to existing machines but must be ordered with new Prima stitchers.

As we go to press there have been a few installations in Europe, and the first system is scheduled for U.S. installation late in 1996.

# Chapter 16

# Commercial Service Providers

A small number of commercial printers and prepress shops have developed new products and services to allow them to deliver on-demand, digital, and customized printing and publishing services to their clients. Besides the advantages already discussed, it is also consistent with corporate America's new battle cry for just-in-time services, quality improvements, reengineered processes, and more timely product deliveries. This new breed of innovators that fulfills the corporate quest for just-in-time manufacturing are employing a multitude of unique digital printing technologies to quickly and economically print in smaller quantities.

They are sculpting a new printing segment — called everything from on-demand printing to digital demand imaging — that, because it targets runs under 5,000, is certain to become a very important part of the commercial segment. Up to now, on-demand reproduction has been mainly the territory of in-plant and copy shops.

The benefits for print buyers are substantial. Printers using digital methods allow companies to get their products to market faster, make last-minute revisions, and reduce or eliminate costly warehousing. And, because of the ability to print directly from digital data, digital printing systems hold promise for the creation of personalized printed materials — a new class of custom-printed products — allowing them to be used in targeted marketing campaigns.

"For years, print has been at a loss to compete with electronic media because it is the same on every page," says Mary Lee Schneider, former marketing director for the Digital Division at R.R. Donnelley & Sons. "Digital printing will allow print to compete with more immediate channels of information delivery while retaining the unique benefits of print, such as its portability."

Personalization is part of the bigger on-demand vision that many hold, but so is distribution. Both may one day be as integral to the process as printing itself. Recent technological advancements, for example, from the debut of several digital color presses to the refinement of high-speed electronic delivery systems, can make on-demand printing in full-color on a global distributed basis a possibility in the near term. But who will adopt them?

Because of the ease of the technology and the potential size of the market, on-demand is attracting many players. Large corporations, for instance, have long exploited the economies of digital printing for office printing.

Now, commercial printers, color trade shops, and service bureaus, because of their expertise in printing and strength in digital color prepress, find themselves in an excellent position to offer streamlined short-run printing to customers.

## Banta Printing

One of the first printers to identify this as a viable market was Banta Corporation. Banta's Demand Documents facility (Eden Prairie, Minn.) was established in August 1991 primarily to offer its computer hardware and software clients the information flexibility they require for their documentation manuals. Utilizing Xerox DocuTechs over three shifts per day, the company was provided with fast turnaround to meet customers' optimum speed.

In an early article on this subject in *American Printer* (September 1992), Chip Fuhrmann, vice president of marketing for Banta's Information Service Group (Minneapolis), said that software publishers often view achieving optimum inventory levels as a double-edged sword. "Software publishers want to have a product available to sell at any given time, yet they strive to avoid excess inventory," he notes. "Obsolescence can be anywhere from 10% to 25% of a publisher's product cost."

"With one specific customer," he continues, "we run weekly production of their manuals. Using our Demand Document solution, we receive orders on Monday, and two days later, the books are printed, assembled, and shipped. A week later, they are in the customer's distribution channel, with zero inventory. This is a concept that software publishers have been striving to achieve for some time."

According to Fuhrmann, publishers' electronic files can be used to generate finished books printed on-demand. A media server is used to interpret those files and feed digital information directly to the DocuTech system. When larger quantities are needed, the firm outputs the electronic files as paginated film, and prints the pieces on more traditional web or sheetfed presses at its adjoining Viking Press facility.

Book turnaround efficiency has been increased further still with the recent introduction of the Signature Booklet Maker for the DocuTech. The system takes a printed set in-line, accumulating, stitching, folding, and trimming it into a booklet. According to Bob Polhemus, manager of integration requirements, systems reprographics for Xerox Corporation (Rochester, N.Y.), the unit can bind as many as 64 pages (16 duplexed sheets) into a booklet.

## Donnelley's Digital Division

Another early adopter of these technologies was R.R. Donnelley & Sons Co., one of the largest printers in the world. For over five years, Donnelley has had a "Books On-demand" facility based in Harrisburg, Va. According to Mark Fleming, former manager of demand services business development, "It is one of a worldwide network of on-demand production operations combining printing with other services, such as maintaining a database of publishers' information, helping customers develop a library of scanned or PostScript files, and directly fulfilling orders to end users."

After months of rumor and innuendo about Donnelley's new on-demand products and services, Rory Cowan, executive vice-president for R.R. Donnelley and Sons, made the first official announcement at the "Hot New Products" session at the Seybold Fall 1994 show in San Francisco. Cowan announced two new Donnelley initiatives: a database publishing service called "PowerBase" and a facility called the Digital Division that will offer a variety of on-demand printing services.

The Digital Division opened its first POD facility in Memphis, Tenn., in the first quarter of 1995. The Memphis location can take advantage of the region's extensive distribution capabilities that include a Fed Ex hub, USPS bulk processing center, 200 trucking companies, and six rail lines. Located next to Fed Ex, the facility will be capable of placing material in the Fed Ex pipeline until the late evening according to Mary Lee Schneider, former director of marketing for the Digital Division.

Schneider says the service is aimed at customers with three characteristics — those who want highly customized materials targeted at specific readers, those who carry a lot of inventory, and those whose materials are in risk of becoming obsolete before they reach the user. One target will be marketing and product literature for major corporations, "but it's really open to anybody who needs that repetitive fulfillment of short-run customized work," Schneider says.

The Memphis facility will offer short-run color output on Heidelberg GTO-DI, Xeikon DCP-1, and Indigo E-Print 1000 presses. For short-run black-and-white output, Xerox DocuTech systems will be used. The facility will also provide binding, fulfillment, and mailing services. At first, the division will offer printing services through a single facility in Memphis, but Donnelley plans to add additional locations by the end of 1995.

## Applications

Catalogs, magazine reprints, and books are all possible applications for on-demand printing. "Matching catalog mailing cycles to inventory availability is a thorny problem for merchandisers," explained Schneider. "Our solution allows rapid delivery based on the latest product availability. This solution finally 'closes the loop' by allowing inventory to drive a catalog which in turn drives inventory — all in real time."

If you ever ordered a book, you know it takes weeks to receive it if it's not on the shelf. That's due to costs associated with inventory maintenance, inventory risk, and distribution. Using traditional strategies today, distribution and inventory management can account for as much as 30% of the overall cost of producing a printed product. Since on-demand printing in its best usage can eliminate most of those costs, customers could save almost 30% by eliminating these problems and their associated costs.

## The Four Keys

Four key components are required to achieve the dream of true customized or personalized publishing — database, digital press, high-speed RIP, and transaction software. The database is based on an Oracle relational database. Named PowerBase, it was developed by Donnelley's Database Technology Services Division and has two information management methods.

The first method stores customer product information in a structured format and is output in a SGML (Standard Generalized Markup Language) format. SGML was introduced in the 1970s as a solution to the different document formats of the proprietary typesetting systems. It comprises a single set of rules that specifies the structure of a document, independent of its format. Prices, SKU identification numbers, and product descriptions are all examples of the kind of data that would occupy separate data fields.

The option of SGML tagging permits the rapid assembly of data for print production projects. The same information could also find its way into an online system used by customer service or sales representatives, according to Ron Brumback, general manager of the Database Technical Services division.

Donnelley can offer up to 125 different catalog styles to manufacturing firms who typically create catalogs "if it's a reasonable structured catalog," Brumback says. These preformatted catalog styles permit the rapid assembly of both general purpose and target catalogs. CD-ROM catalogs can also be created.

The second PowerBase method stores and retrieves unstructured information such as images, documents, audio, or video clips. Customers can retrieve text or graphics files for incorporation into their own QuarkXPress or PageMaker page layouts, Brumback says. With PowerBase's OPI capability, users can download low-resolution, FPO (for-position-only) versions of scanned images that are replaced with the high-resolution images during output.

Donnelley's new Digital Division integrates this second PowerBase method into its rapid print services. The idea is that companies can store product information in a central location, from which they can generate catalogs, product literature, online customer service references, and other marketing-oriented material in print or electronic form. In most cases, the information is stored, maintained, and printed at a Donnelley location. Ultimately, customers will transmit their documents to the facility from their own locations. The PODs will be networked to R.R. Donnelley customers and other facilities with a high-speed, frame-relay network.

## The Ripping Problem

Of the first three keys, the PowerBase database is in place, and digital presses are commercially available. An unresolved issue is the RIP. At services bureaus, new RIPs are notorious for causing problems. That's because the horsepower required for ripping is tremendous, and RIPs are very complicated computer hardware and software. There was a period of time, for example, when the Apple LaserWriter printer was the most powerful computer you could buy from Apple.

The capabilities of a RIP to incorporate changes to a page (i.e., customize) and have it printed at print engine speeds has been controversial since the first digital presses were shown at IPEX. At different shows, manufacturers would give different answers to the same questions about the area and complexity of the page that could be customized.

Today, it appears that the current RIP technology will allow either a small (4×5-in.), four-color area or a letter-size area of the black plate alone. At Spring 95 Seybold, Donnelley announced it was beta testing a new RIP from Barco that could allow four-color customization of a full letter-size page. Historically, RIPs are difficult and time-consuming to debug. However, according to Donnelley, the system is "up and running with no problems and producing live work."

## Transactional Costs

One issue that develops with this new print model involves transaction costs. If it is true that long-run print will segment into multiple segments of short runs, then transaction costs will rise dramatically. Assume it costs you, as a print provider, $100 to process an order for a long-run print job, from a sales and administrative point of view. If it costs the same amount for multiple short runs, then you will develop financial problems because it will cost more to process more jobs. This issue is only beginning to become evident.

The way Donnelley is dealing with this issue is by giving the ordering power to the customer. Using customized software called "Order-It™" the clients make their own orders, which also results in an invoice and a debit. According to Mary Lee Schneider, "Digital printing is about process reengineering, from front to back. You can't just lay digital print over a traditional model."

## Questions

Why would Donnelley go for a centralized rather than a distributed approach? With their stake in Alphagraphics Print Shops of the Future, they could install a digital press in major U.S. markets for local printout. We think that centralization gives them several advantages:

1. There are many different printing systems in one place so customers can have their choice of quality, performance, and cost. Local installation might be limited to one system.
2. Prepress functions can be centralized with a lot of talent in one place. Would each local site have such talent?
3. Bindery, the missing link in digital printing, can be centralized as well. Bindery equipment is quite expensive. As explained in the chapter on the bindery, it has created the new challenge in digital printing.
4. Federal Express rates may not be higher than the cost of a messenger to go across town to pick up the printing from the local service.

## Moore

Moore Corporation, best known for its business forms printing and systems, has begun to recognize the growing demand for color printing in documents of all kinds as a potential growth area. Thus, in April 1993 it formed a strategic business unit called Moore Graphic Services to provide commercial color printed documents to its traditional customer base. Before opening this division, Moore had been handling commercial print runs for its customers through partnerships with other companies. The advent of digital color printing was an important part of its decision to move color printing in-house and to provide its customers with a fuller range of printing services.

It is currently evaluating several technologies for quick-turnaround digital color printing, including digital presses manufactured by Indigo and Xeikon. Under Moore's new strategy, which it calls "print management," the company provides complete services for its customers. It recognizes that for every $1 a customer spends on forms, it typically spends $5 on other commercial printing needs. One area it will emphasize is small-quantity, quick-turnaround product offerings. The company recognizes the need for some customer education regarding the benefits of short-run demand printing versus the traditional approach of warehousing large quantities of documents.

After months of speculation about what Moore will do with the digital presses, their announcement at the Spring 95 Seybold show quieted the rumor mill. Not totally unexpected, Moore made an announcement not unlike the announcement made by Donnelley at the Fall 94 Seybold show, about their entry into the digital printing market. Called the Moore Digital Print Management Series, it encompasses a global network for creating, distributing, and printing documents.

According to George Gilmore, president of Moore Business Systems, "This announcement reflects our ability to deliver the entire range of information handling products and services our customers need. Our new capabilities allow us to satisfy our customers needs for shorter run lengths and faster turnaround times." Gilmore is one of many senior managers who have recently joined Moore.

Gilmore also said, "Today, Moore is able to deliver a workflow-process-enabling technology for on-demand digital color and black-and-white printing." The workflow-enabling technology mentioned is several software packages created by Moore that aids in the creation and preflighting of PostScript files. The Digital BrochureShop software contains a set of templates designed to help customers create promotional literature quickly. The Digital Pilot software is designed to overcome printing problems.

A third piece of software, called Digital ColorQuick, is designed to overcome the most difficult and fascinating aspect of digital printing known as variable imaging, customization, or personalization. According to Moore, the company's services are targeted at three types of customers: those who want highly customized materials targeted at specific readers; those who carry a lot of inventory; and those whose materials are at risk of becoming obsolete before they reach the user.

These documents could include custom posters, new insurance offerings, or targeted mailings to the demographics of the individual.

To accomplish this task, a database has to work with text and images, and the digital press requires a fast RIP (raster image processor). Moore's Digital InfoBase is the database that houses the images, text, and templates. It allows customers to request and build documents on-demand. Digital ColorQuick is the hardware and software that enables the high-speed ripping of variable color or monochrome images, variable data, fonts, and document templates.

According to Moore, the variable imaging can be accomplished automatically—there is no need to slow production while the print engine waits for the database or RIP, and Moore's system is five times faster than any technology available today. The speed claims are hard to believe, but these claims can be accomplished if they rip files to a disk first or RAM (random-access memory) and then download the ripped pages over a fast computer bus.

The final piece of the Moore puzzle is a global, digital, frame-relay network. It will allow Moore's customers to send, receive, and distribute documents worldwide. The worldwide ability is enhanced by Moore Multicopy, a wholly owned European franchise, with relationships in Asia with Toppan Moore.

# Section III

# Technology

# Chapter 17

# All Technologies

Although there are a host of technologies used to print pages, the vast majority of pages today are created using one of three technologies: offset lithography, dry offset, and electrophotography. Offset lithography is used both in duplicators at the 11×17-in. size and below as well as in presses at the signature-level or 17×22-in. size and above. Dry offset technologies are the same as offset except that fountain solution is not needed; i.e., there is no concern for ink/water balance.

The third technology is electrophotography, or copying technology. This technology uses electrostatic charges to build an image and transfer toner to the page where it is finally fused to the paper.

The two offset processes are used for high-quality reproduction, especially where color is concerned. However, a major advantage of the printing process is the ability to print several pages at the same time or a signature consisting of 8–16 or more pages at one time. Combined with automated bindery equipment, this is the most efficient approach to the production of publications with large page counts (100 pages or more), reproduced in high numbers (1,000 copies or more).

| | *Best for* | *OK for* | *Worst for* |
|---|---|---|---|
| **Offset lithography** | High-page-count, high-copy-count documents | Moderate-page-count, moderate-copy-count documents | Short runs of any kind of work |
| | Color reproduction | | |
| | Special or heavy paper stocks | | |
| **Dry offset** | Very high quality halftones, color | | |
| **Electrophotography** | Short-run reproduction | Limited-run documents of moderate page count | High-quality color |
| | Most black-and-white flat forms | | |
| | Jobs where information is in electronic form | | |

## Overview of All Technologies

Our definition of on-demand printing does not talk about any particular technology. It describes on-demand in terms of short notice, quick turnaround, and economical short runs. We realize that this is not the common definition. We came to this definition after talking to many people who have used various technologies to accomplish on-demand printing and publishing.

The technologies in these devices as well as their origins are quite varied. The Xeikon and Chromapress electrophotographic-based technology emerged from the laser printing market (see Chromapress chapter). The DocuTech and Lionheart use electrophotographic technology and combine it with LED technology, which came from the copier market.

Some technologies are hybrids of two other technologies. For example the Indigo combines electrophotographic technology with traditional printing technology (offset press with blankets and impression cylinders). The Scitex press is another hybrid combining high-speed inkjet technology developed at the Kodak Dayton, high-speed inkjet facility (which Scitex bought patents and all) with high-quality inkjet technology from Scitex's IRIS inkjet division. Another hybrid, discussed in Chapter 30, is the ability to install Scitex inkjet personalization technology at the end of a commercial press.

And we have run into commercial printers who have found other tools and procedures to deliver demand printed products.

Since, clearly, no one technology is associated with on-demand printing, it is difficult to predict what other technologies are well suited for on-demand printing. Therefore we have decided to review all printing technologies that could be used alone or in combination for other on-demand solutions. After reviewing all the technologies we will make some bold predictions about which technologies hold the greatest promise as future on-demand technologies.

We can break down output technologies into two broad groups: impact and nonimpact technologies. If you've been following laser printing or color proofing technologies, which are nonimpact technologies, you might wonder why we even discuss impact technologies. The reason is that the workgroup market, or data processing market, uses high-speed impact printers.

## Impact Printers

Basically an impact printer uses a printing mechanism that comes in contact with another element before reaching the page. The first impact printers were typewriters, which were followed by dot-matrix printers. Obviously typewriters and dot-matrix printers are not appropriate for on-demand applications, but you might be surprised to know that high-speed impact printers are still used in data processing facilities.

Impact printers today still perform a lot of the grunt work in data processing shops. Despite the "nay-sayers" who predict their demise at the hands of high-speed nonimpact technologies, such as electrophotographic-based technology, they continue to survive. Although competing technologies have captured significant portions of the industry's printing needs, the impact printer remains the choice for companies producing large volumes of printed material that can tolerate characters of slightly lesser quality.

These printers can cost $35,000 or more. They are rated in abilities such as how many pins are in the printing head and how many lines can be printed simultaneously (i.e., 33-pin print head capable of printing three lines of text), how long the print head life is (i.e., 1 billion characters), and what output languages can they accept or emulate.

## Nonimpact Printers

Basically nonimpact printers are printers that do not strike a hammer to a platen as typewriters do. Usually a heat-sensitive paper or laser/copier type technology is involved.

## Electrophotography

Also called electronic, electrostatic, or xerography, electrophotography is the underlying technology in copiers and laser printers. This technology uses a drum charged with a high voltage and an image source, often from a laser or LED (light-emitting diode). The image source paints a negative copy of the image to be printed onto the drum.

Both plain-paper and special-paper copiers use electrophotographic coatings of selenium, zinc oxide, or cadmium sulfate to hold the electrostatic charges in the dark or imaged areas. When the light falls onto the drum, the drum is discharged. The toner, which some call "dry ink," adheres to the charged portion of the drum. The drum then fuses the image onto the paper by pressure and heat. Electrophotographic printers use electrostatic charges in the imaging process.

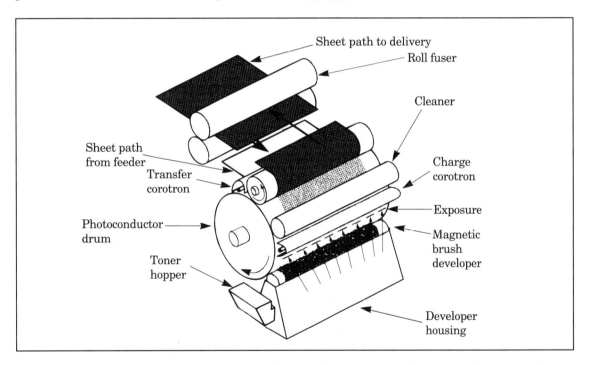

A typical sheetfed electrophotographic process. Photoconductor drums or belts carry the image and toner for transfer to the paper.

This is a popular technology, but it is subject to three serious deficiencies:

**1.** Charge voltage decay—first between the time of charging the photoconductor and second during exposing and toning—can affect image density and tone reproduction as the amount of toner transferred to the image is dependent on the exact voltage of the charge on the image at the instant of toner transfer.

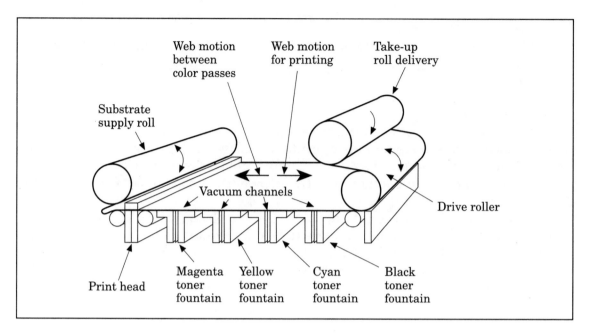

A multicolor electrophotographic process. A webfed approach is shown as well. Each print head transfers a different color, and the combination of all of them produces full color.
*Courtesy Versatec*

2. Toner chemistry is not completely understood, which can cause variations in batches of the same toner and high cost for formulating special toners.
3. In liquid toner systems, the isopar used to disperse the toner is a volatile organic compound (VOC), which might require venting and is subject to environmental regulations.

## Field Effect Imaging

Field Effect Imaging is a high-speed, high-quality color printing process that uses variable-density pixels with 500-dpi resolution. It uses three novel materials.

The new materials are: *X1,* which is a thin-film dielectric ultra-hard writing surface onto which electrical charges are deposited by an ***M-Tunnel*** write-head to generate powerful electrostatic fields across the thin-film insulator, which fields cause the pick-up of *X2* "ink bites" whose thickness is proportional to the strength of the fields. The bites of ink, in pixel size, are carried by the X1 surface to a print position where the ink is transferred totally to the print substrate, which can be either paper, plastic, cloth, or metal. See more discussion on XMX and its technology in the chapter on "Futures."

## Inkjet Printing

Inkjet printing is a digital printing system that produces images directly on paper from digital data without a press-like imaging machine — that is one that requires an image carrier. These inkjet printing systems use streams of very fine drops of dyes that are controlled by digital signals to produce images on plain- or special-paper surfaces.

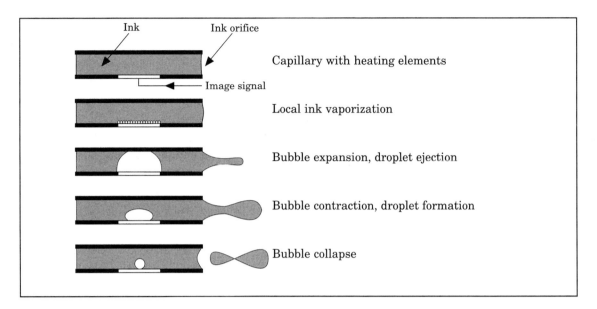

Thermal inkjet formation.
*Courtesy of SPSE: The Society for Imaging Science and Technology*

There are three types of inkjet printers:
- Continuous-drop
- Drop-on-demand
- Bubble-jet

Some have single jets, and others use multiple jets. Most inkjet printing is single- or spot-color printing of variable information like addressing, coding, personalized computer letters, and other direct mail advertising. Inkjet systems are used for color proofing but have been too slow for short-run color printing. However, this will change.

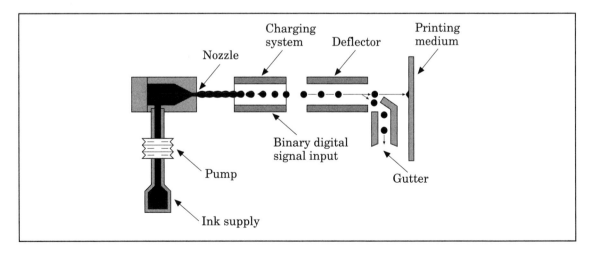

Binary continuous inkjet technology.
*Courtesy of Iris Graphics*

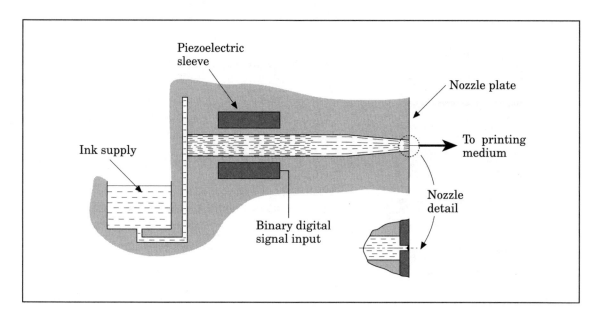

Drop-on-demand inkjet technology.
*Courtesy Wolfgang Wehl, Siemens AG, and reprinted by permission of SPSE: The Society for Imaging Science and Technology*

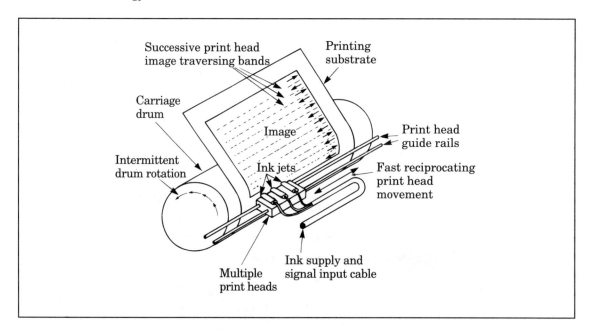

An inkjet positioning and paper feed mechanism.

## Ionography

Ionography is a technology that uses ion deposition or electron-charge deposition printing. It is similar to direct electrostatic, except that in this type of indirect electrostatic, the image is formed on a dielectric surface and then transferred to plain paper. This process was developed by Dennison Manufacturing Co. and is called Delphax. The

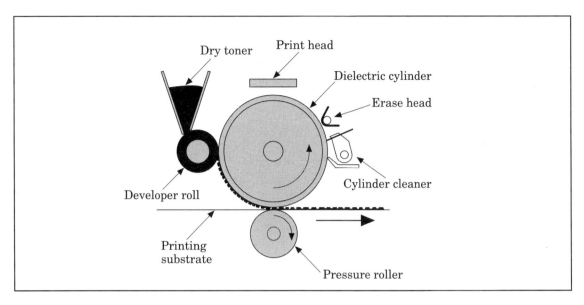

An ionographic engine.
*Courtesy Dennison Manufacturing Co.*

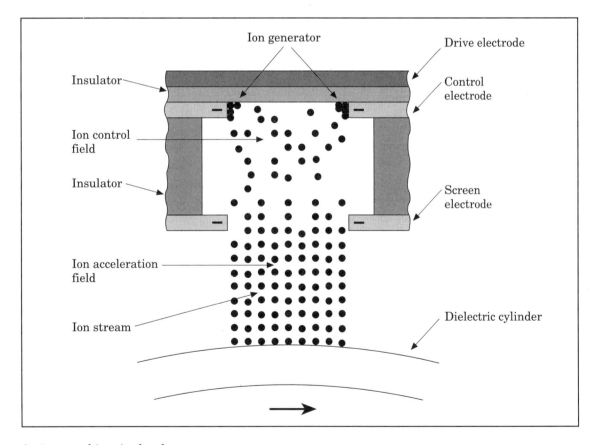

An ionographic print head.
*Courtesy of SPSE: The Society for Imaging Science and Technology*

image is produced by negative charges from an electron cartridge onto a heated dielectric surface of aluminum oxide using a special magnetic toner.

It is used only for single- or spot-color printing because the pressure of image transfer and cold fusion fixing of the toner can distort the substrate. Systems are in use for volume and variable printing of invoices, reports, manuals, forms, letters, and proposals and for specialty printing of tags, tickets, and checks.

## LED Array Imaging

An array of light-emitting diodes (LED) the width of the page can be used to image a page. Generally the LEDs don't move, which allows all the pixels in one or several rows to be imaged at the same time. They are used in some of the high-speed copiers and on-demand digital presses.

## Magnetography

Magnetography is a nonimpact printing technology like ion deposition except that a magnetic drum is used. The magnetic image is created by a set of recording heads positioned across a magnetic drum. A monocomponent magnetic toner is applied to the drum to develop the image. It is transferred to paper by light pressure and an electrostatic field. The toner is then fused by heat.

Ironically, spot colors can be used but not process colors because the toners are dark and opaque. The systems are used for printing business forms, direct mail, lottery tickets, numbering, tags, labels, and bar codes. The Bull engine is used almost exclusively for these systems.

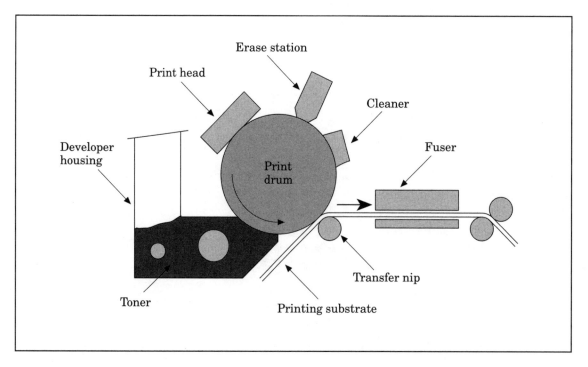

A magnetographic print engine.
*Courtesy Bull, S.A.*

## Thermal Printing

In thermal printing, dots are burned onto specially coated paper that turns black or blue when heat is applied to it. A line of heat elements forms a dot-matrix image as the paper is passed across it, or a serial head with heating elements is passed across the paper.

## Thermal Transfer Printing

A thermal transfer printer uses digital data to drive a thermal print head to melt spots of dry ink on a donor ribbon and transfer them to a receiver. The technology was introduced in 1970, but printers were not available commercially until the mid 1980s.

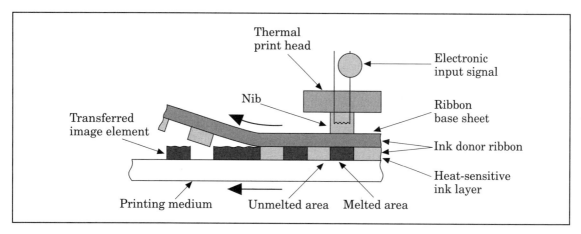

Thermal transfer printing.
*Courtesy of Journal of Applied Photographic Engineering, vol 7, 1981 by SPSE*

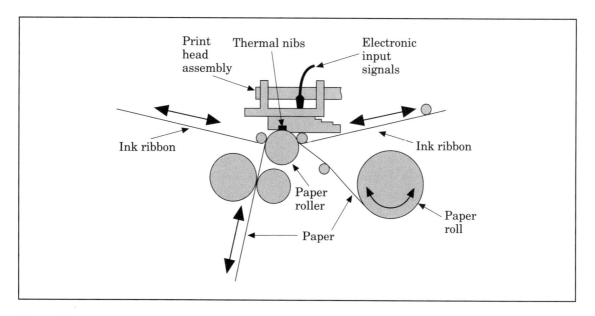

Thermal transfer print engine.
*Courtesy Seiko Instruments USA, Inc.*

They are used for word processing, computer output printing, facsimile, graphic and pictorial color printing, labels, and other applications in single or multicolor. When the solid ink on the donor is replaced by a sublimable dye, the thermal head converts the dye spots to gas spots that condense on the receiver. This configuration of the printer is the thermal transfer dye sublimation engine used for color proofing.

# Chapter 18

# Copier Technology

## Copier History

In 1938, the first electrophotographic image was produced by Chester Carlson, a physicist and patent attorney. Like many other great inventions, financial and market success did not happen overnight. In fact, it was not until 1944 that Carlson persuaded Battelle Memorial Institute to conduct additional experiments with his process.

During the period from 1944 to 1948, the basic process and technology were refined, and from 1944 to 1946 Battelle invested its own resources. But things changed in 1946. The Haloid Company (which eventually became Xerox Corporation) of Rochester, N.Y., was granted a license to develop electrophotography. Later, in mid-1948, the U.S. Army Signal Corps sponsored research at both Battelle and Xerox, intending to apply the process to military photographic areas. The Xerox Corporation's first commercial product based on electrophotography (now renamed "xerography," from the Greek words for dry writing) was a hand-operated copying system that appeared in 1950.

In 1960, the first automatic office copier — the Xerox 914 — appeared on the market. At six copies per minute, the 914 had most of the capabilities of today's "personal copiers." In 1970, IBM marketed the Copier, which featured an innovative, reusable, organic photoreceptor. Five years later, the Eastman Kodak Company, a company which already had a century of experience in the imaging field, joined the electrophotographic copier industry with its Ektaprint series. The latter system incorporated a re-circulating feeder for originals and an online binder/stapler. It was the beginning of demand printing.

From the industry's beginnings, many United States companies developed products to take advantage of growth in the market. All but a few of these have abandoned their efforts in the face of Japanese competition.

## Color Electrophotography

Color electrophotography had its commercial beginnings in 1969 with 3M's Color-in-Color copier, which used electrophotography to form a mask that controlled the migration of colorants. In 1973 Xerox released the 6500 Color Copier followed by the 1005 model, and Canon followed in 1978 with the Canon T machine.

Eastman Kodak entered the color copier arena in 1988 with the ColorEdge copier. At 23 three-color copies per minute (cpm) and 70 cpm for single-color copies, the Color-

Edge represented some technological innovations in electrophotography. The Color-Edge used a flexible photoreceptor belt architecture. The original was placed on the platen and a flash exposure through each of three filters — red, green, and blue — performed the color separation. Development of the electrostatic image was via a magnetic brush. Kodak used the transfer-directly-to-paper technique to synthesize the image from the three colorants. The last toner transferred is yellow, and then the image is forwarded to the heated roll fuser. The ColorEdge could not copy in black only.

The Xerox 1005 is the oldest of the full-color copiers considered here. Its genesis is the Fuji-Xerox 6800, a second-generation of the original Xerox 6500. The 1005 was a drum copier with a scanning optical system that performed the color separation function via the three-filter exposure technique. It had the usual three development stations around the periphery of the photoreceptor drum. Transfer was directly to the paper on an intermediate transfer roller.

The Canon CLC (Color Laser Copier) had four toners, including black, and used digital scanning and printing approaches. Color copying and printing were truly born with the CLC.

## Other Color Printers

In 1987, Mead Imaging announced its new color imaging technology Cycolor. Cycolor technology represents nearly a decade of development at the Mead Corporation and its subsidiary, Mead Imaging. The Cycolor system was the first practical non-silver-halide, dry-process, high-resolution, continuous-tone color imaging technology. It was based on microencapsulation, the same technology used in carbonless paper.

Cycolor was a two-component imaging system consisting of (1) the film or donor sheet and (2) the receiver sheet. The Cycolor film is coated with billions of cyliths, each only several micrometers in diameter. First, the film is exposed to the colored light. Those cyliths sensitive to the colored light become hardened, while the un-exposed cyliths remain soft. The donor is then pressed against the receiver sheet by the pressure rollers.

The soft cyliths burst under pressure, transferring their dye to the receiver. The dye precursor from the burst cyliths then reacts with chemicals on the receiver sheet, forming the appropriate color. Continuous-tone capability was possible, since cyliths harden in proportion to the amount of the light exposure.

In terms of color printers, in 1985 Howtek demonstrated the HR-1 "Thermo-Jet" solid inkjet printer. Exxon and Dataproducts announced an agreement to cooperate in the commercialization of Exxon's solid inkjet technology. Named E/D Venture, the agreement resulted in a monochrome office printer called the SI (for Solid Ink) 480 by Dataproducts in 1986. Howtek continued to develop its color solid inkjet printer, renamed the Pixelmaster.

The first thermal dye transfer printer, the Sony Mavigraph, was shown as a proto-type in 1982, the same year that Shinko Electric and Seiko introduced the first conventional wax thermal transfer printers. The Mavigraph did not become commer-cially available until the spring of 1986. Only within the last decade has color print-ing technology developed the methods and processes to make pictorial images at the quality, speed, and cost that users can justify.

## High-Speed Black-and-White Copiers

In 1989 Xerox announced its DocuTech 600-dpi laser printer capable of printing 135 pages per minute (8100 pages per hour) in a format size up to 11×17 in. About the same time Kodak announced its 1392 Lionheart 300-dpi, 92-ppm (5,520-pph) PostScript-based electronic printer. These electronic printers are used for short-run variable-information printing and the new on-demand publishing market. Their speed and resolution ushered in a new age of digital printers that competed with offset printing.

Although generally referred to as on-demand presses, the Xerox DocuTech and Kodak Lionheart use electrophotographic technology, which allows us to categorize them in the same category as high-speed copiers. On the other hand, some of the paper handling and high-speed synchronization make them much more than simple copiers. Each of these units is discussed in more detail in later chapters.

These units are designed to print pages fast. Units in this segment can produce between 150,000 and 400,000 copies per month and range in price from $25,000 to $40,000.

If you are considering a high-speed device, then you are most likely considering an expensive investment ($200,000–400,000). For devices in this category, there are certain features to choose from. Of course, the amount you pay determines how many of these features you get.

For example, all machines can duplex, which means it can print on two sides. A basic duplexing feature would allow paper storage of 50 letter-sized sheets or output of 100 double-sided pages. The more advanced machines could make double-sided 11×17-in. pages. And a number of newer machines offer "one-pass" technology, which the manufacturers claim will result in fewer paper jams.

Stapling is a function grouping under finishing services. Usually this is available on both the sorter and OCT configured machines. Occasionally it is standard, but usually it is an expensive option. The more elaborate stapling options can cost up to $3,000 and will give you a choice as to where the staple is to be placed. The basic stapling feature will put a staple in the upper left-hand corner, while more sophisticated designs allow users to position the staple in up to three different places.

Another feature of these devices is the reduction and enlargement option. Since this feature can be found in most devices, the only differentiating aspect is the specific percentages — 10%, 50%, 90%. More sophisticated units allow reduction or enlargement in increments of 1%.

The copy editing feature allows users to make changes to the copy. More elaborate systems allow you to cut and paste text, while simple systems allow for the blocking out of certain areas of text.

More expensive devices come with a graphical user interface and a TV-style monitor. Some even offer this option in touch-pad versions. The photo-mode option is an important feature for picture work, because this option allows users to darken or lighten the printing of photos. If you ever need to copy books, an important feature is the book copying feature. The book copying feature lets users place a book in the open position and individually copy both the left- and the right-hand pages in a single run.

For complicated jobs, a job programming feature allows users to program the copier to handle the complete job with just the pressing of a few buttons. The job ticket function of the DocuTech is a good example of this function.

## Issues with Copier Technology

Most electrophotographic products transfer the image to some intermediate drum, roller, or belt to collect and store the sequence of single-color images. In some cases, the paper is fastened to these intermediate devices and the image is transferred directly to the paper. To avoid misregistration, precise positional control of the transfer intermediate and the photoreceptor image is absolutely essential. Large, uniform color areas necessitate uniformly applied toner layers.

Achieving spatial uniformity requires spatial precision of all electrophotographic process steps. The colored toners are transparent. Any variation in the thickness of this transparent layer results in a color variation. Overprints (one layer of toner on top of another) that are not uniform exacerbate color variations. With black-and-white electrophotography, toner is virtually opaque, and sufficient quantity is put down to make the image black.

Provided there is enough toner to make the image black, there is essentially no sensitivity to thickness variation. Uniformity of a toner layer over the duration of the run is also needed if the first image is to have the same color as the last image. This places a different, but equally demanding, set of conditions on the dynamic performance of the electrophotographic imaging system elements.

To reproduce a large number of colors from various sources, a large color gamut (color volume) and palette are desirable. Achieving large gamuts necessitates transparent toners. If the toners are not transparent (assume for a moment that they are completely opaque), the most visible color would be the color of the toner on top. The toner transparency issue becomes especially significant when making overhead transparencies for projection.

A related issue, light scatter, is also important in this application. If the toner layer is at all light-scattering, the color image on the screen looks gray and less colorful (low chroma). Although the toners themselves contribute

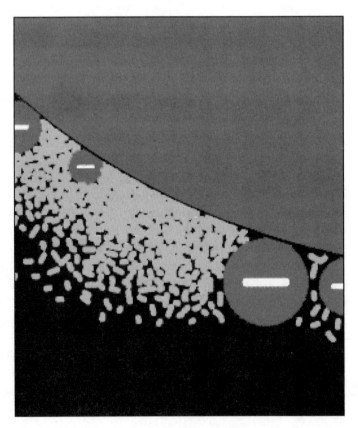

*Courtesy of Agfa, a division of Bayer USA*

significantly to scattering, fusing methods can also alter the projected color by controlling the surface characteristics of the overhead projection.

Colored toners are particulate, varying in size from about 5 micrometers to 20 micrometers in diameter. The fact that they are particles causes the image to have some inherent "graininess," or a very fine, structured lightness variation of solid color areas. Smaller-sized toners produce lower graininess, which is one reason that the high-quality graphic arts electrophotographic color proofing systems use very small particle toners in conjunction with liquid development. Liquid development is a convenient method of applying small toner particles to the surface of the photoreceptor, because smaller, dry powder toner particles are hard to control.

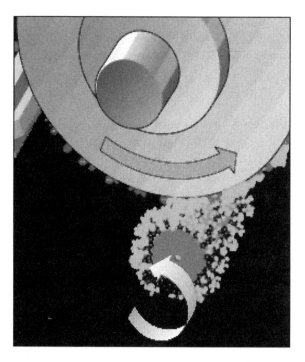

*Courtesy of Agfa, a division of Bayer USA*

## Toner Coverage

Black-and-white copiers and laser printers are designed to develop text images with about 5–15% of the page area covered. Heavy monochrome graphics applications can raise this requirement to about 25–30% total page coverage.

For good print quality over the entire length of a run, the electrophotographic development subsystem must be able to apply toner to the photoreceptor at these points of coverage. The situation is different for color.

## Color Is More Difficult

With color electrophotography there are four toner layers, and therefore four times the "regular" (black-and-white) rate of toner consumption. Secondly, for pictorial or graphics images (which represent the bulk of color imaging at present), larger areas of the paper are covered by toner. The world, on average, is gray and has about 20% reflectance, which translates into 80% area coverage for each of the three colored toners.

Comparing this to the black-and-white text case, toner consumption for color pictorial and graphics applications will be about 10–15 times more. The unit image consumables cost will rise considerably, due to the expensive colorants used and the increased toner consumption. This also highlights the advantage in printing those images that contain no color with a black toner, and using black where possible in colored areas. The lightfastness and waterfastness of toned images has not been a significant issue. Dyes are becoming options for toner manufacturers, and because dyes fade, we may experience greater light deterioration of toned images than we have seen in the past.

Toner is a plastic polymer and is largely immune to the effects of water: it is extremely waterfast. Early electrophotographic copiers tended to use large amounts of toner, which gave the copy a plastic-coated look. Additionally, when a color copy was folded, the toner would often crack and flake. This problem is diminishing with newer toner technologies.

## Toning

There are several popular options for toning or developing the electrostatic image to make it visible. These are:

- Single-component development versus dual-component development
- Liquid development versus dry power development

The most common method of applying toner to the photoreceptor is the "magnetic brush." The magnetic brush is formed by small bits of ferromagnetic material (iron) under the influence of a magnet (the magnetic field).

The filings form in lines on the paper, showing how the magnetic field goes from the North to the South magnetic pole. If magnets are arranged inside a cylinder, for instance, and some form of iron compound, in particle form, is placed around it, a magnetic brush will be formed. In electrophotography, the iron compound that is responsible for the formation of the brush is called the carrier.

The carrier, also called the developer, is one component of the development system. Toner, the second component of the system, is responsible for forming the image on the copy. Single-component development systems have the iron compound and the toner combined into one; hence the term "single-component."

Subtractive-color toners must be transparent in order to form colors. Single-component toners that use magnetic means to get the toner to the photoreceptor surface suffer some drawbacks when applied to color, because the magnetic material used in black toner is not transparent to light. Therefore it is extremely difficult to generate single-component toners that yield a color gamut adequate to reproduce a wide range of colors.

If some transparent magnetic material were developed, or if other-than-magnetic means were used, single-component development systems would be viable contenders for producing process color. Single-component developers with opaque toners are now used with spot-color copiers or printers.

## Liquid/Powder Toner Development

A key factor in choosing to use liquid or powder toner is the particle size. Applications that require the highest image quality (for example, graphic arts color proofing) require small toner particles for quality. Small toner particles are difficult to deliver to the photoreceptor surface with a conventional magnetic brush. An additional factor is that if the small particles become airborne, they are difficult to control and tend to land in all the wrong places.

Suspending these particles in some liquid provides an alternative, particularly if the application requires small-size toners. The major difficulties are the control of the liquid carrier, which must be recovered or trapped to prevent contamination of the components of the copier and the surrounding environment.

Once the electrostatic image is toned it must be removed, or transferred, from the photoreceptor. Two approaches are in use in color electrophotographic devices:
- Sequential transfer to paper wrapped around an intermediate transfer roller
- Transfer to an intermediate belt

The first technique is called "multipass," and the latter scheme is called "single-pass." These terms really refer to the number of passes the paper makes past the transfer station. The multipass approach has the advantage that there is one less transfer station, which helps decrease cost and increase reliability.

On the minus side, using paper as an intermediate (as opposed to some material with known, constant properties) offers opportunity for the misregistration of the cyan, magenta, and yellow toner images. Paper is noted for its variable properties, particularly its dimensional changes with water content.

Also, transferring directly to the paper may complicate the paper path and create opportunities for misfeeds and paper jams, decreasing system performance. Implementation of the single-pass method dictates intermediate transfer of the four colored images, and then one final transfer step to the final substrate. The obvious advantages are the simple paper path, with its attendant reliability, and the more consistent image transfer from the photoreceptor.

One of the major determinants in the choice of "belt versus drum" photoreceptor is the desired speed (expressed in copies per minute) of the device. However, copying speed is not the only factor in the belt versus drum tradeoff. A belt photoreceptor must be flexible. This flexibility offers a chance to make the copier or printer package smaller yet fit all of the necessary components. Drum photoreceptors are a well established technology. The rigid drum eliminates some positioning problems that might cause color misregistration.

Electrophotography is similar to a printing press in that it sequentially forms an image on paper. Black-and-white electrophotography is a high-speed process. Process speeds in excess of two feet per second (120 copies per minute) are possible.

However, because of the sequential nature of color electrophotography, and because a single photoreceptor is used for all colors, speed is reduced by a factor of four. A high-speed device could be configured like a printing press, where each colored toner is imaged at a separate station. Such a system would obviously be large, but it could function at the 120 copies-per-minute rate.

# Chapter 19

# Competing with Copiers

Today most of the on-demand products use electrophotographic or copier-based technology. These include DocuTech, Lionheart, Chromapress, and Xeikon. The question becomes "Will this technology continue to dominate the on-demand market?" New black-and-white and color copiers are incorporating the laser printer technology that will allow them to compete as on-demand systems.

Other technologies can fulfill the same requirements. Sometimes they are in direct competition with electrophotography, sometimes they offer totally distinct features that mark them for different applications. Among the characteristics on which competitive technologies can be compared are speed, hardware cost, cost per copy, ability to do continuous-tone printing, image permanence, and media restrictions.

The three technologies that have the greatest chance of competing are electrostatic, inkjet, and thermal transfer. As you see in the more in-depth description that follows, the toughest technology to differentiate from electrophotographic is electrostatic, because electrophotographic printing uses electrostatic principles. The differences are in the formation of the charge image and the need for special paper.

Inkjet and thermal transfer are easier to distinguish. Inkjet works by spraying one or more colors of ink that are aimed by electric field deflectors (electromagnetically) onto paper. It can favorably compete on a cost-per-image basis and on page-per-minute rates. Permanence of the color image remains a question with water-based inks, and quality is an important issue.

Thermal printers apply heat to a ribbon carrying waxed ink to transfer it to paper. All variants of thermal transfer printing do not offer the printing speed for high-end printing and copying applications. The highest-quality thermal transfer is dye diffusion, due to its continuous nature. However, it is also the most expensive on a unit-image basis.

## Electrostatic (Electrographic)
Equivalent to electrophotography in many respects, electrostatic technology (electrography or dielectric) differs in the method of forming the electrostatic image. Recall that electrophotography discharges, via light, a photoconductor that has been "sprayed" with an electric charge.

Electrography, on the other hand, places a charge on specially coated paper where an image, or dot, is to be formed. A wire "nib" print head, comes in contact with the dielectric coated paper at a suitable voltage, and places a charge on the paper. Four passes of the print head are required to create a full-color image.

In most configurations, the charged image on the paper is toned and then the paper is rewound, and the next image charged, representing the next color, is placed on the paper. This process is repeated until all four colors are printed. The development of the electrostatic image is identical to electrophotography. However, liquid development is the method of choice due to simplicity and cost for wide formats.

Dry toner development would be more difficult and costly over such large widths due to the increased complexities of the magnetic brush development system. Like electrophotography, electrography yields images of high permanence and resistance to light,

*Courtesy of Agfa, a division of Bayer USA*

water, and mechanical abrasion. These images contain toner that has a lower fraction of polymer (plastic) than electrophotography, so the images do not have a plastic-coated look and feel.

## Inkjet

Inkjet printing can be divided into basically two categories: continuous and drop-on-demand (DOD). Continuous inkjet, which had its beginnings in the early 1960s in the recording of electrical signals on a paper chart, employs a continuous stream of fluid. Natural forces cause the continuous stream to randomly break up into little droplets.

The process parallels that of a garden hose: the water first comes out as a stream, but before it hits the ground, drops of various sizes are formed. This random breakup is totally unsatisfactory for printing since the drops can not be placed accurately on the paper surface. A solution to this problem is to "stimulate" breakup by applying a high-frequency pressure variation to the ink stream, causing the drops to form in a known and repeatable manner.

By putting an electrical charge on the drop, at the instant it breaks from the continuous stream, and letting the charged drop fly through an electrical field, the position of the drop on the paper can be precisely controlled. Drops that are not directed or written to the paper are deflected to an ink recirculating system to be used again.

The Iris Graphics inkjet printer is an example of this technology. It forms drops at the rate of 1 million per second and places them at 300 dots per inch. This high drop rate, coupled with multiple nozzles, could enable the technology to increase its printing rate from about 2–3 pages per minute to perhaps 20–30 pages per minute, or more, with one pass. An advantage to be gained from using the continuous inkjet method is that the fluid is not afforded the opportunity to dry in the nozzles.

The drop-on-demand (DOD) inkjet method has been described as the "oil can." A closed chamber with a small nozzle at one end, and filled with ink, is reduced in volume via a piezoelectric actuator. The decrease in chamber volume forces the ink fluid out through the nozzle.

A variation on this theme is the thermal inkjet, or "bubble jet," that, instead of reducing the ink chamber volume, the fluid ink is expanded, by heating a small volume of ink, forcing a drop out the nozzle. The drop ejection rates of the Hewlett-Packard PaintJet thermal inkjet printer are limited to perhaps 20,000 drops/second largely due to the complexities of refilling the ink chamber. Increasing the printing speed of DOD inkjet technology would require a multiple-nozzle or array approach.

A variant within DOD is now hot melt or solid inkjet, which uses a fluid that is solid at room temperature and liquid at higher temperatures. The first color configuration was the Howtek Pixel Printer. It conceptually worked the same as the water-based systems. The difference is that the hot ink solidifies as it hits the paper and does not interact with the paper in the same way as water. The solidified ink is bound to the surface of the paper and does not flow into the paper fibers, resulting in more clearly defined dots and more saturated colors. Hot-melt inkjet is essentially a plain-paper process, which could prove to be a significant advantage.

Most inkjet products, whether continuous or drop-on-demand, use water as the ink fluid to carry the colorant. The water can evaporate at various times and cause nozzle clogging, which plagued early inkjet products. Water, as the fluid base, poses additional challenges when it interacts with paper. Ink bleeding within the paper causes the spot to grow, sometimes in an irregular manner, resulting in poor print quality. For color, the ink-paper interaction problems are accentuated because color formation demands that ink drops be placed on top of each other. This increases the amount of ink that must be absorbed by the paper.

An additional consideration for color printing is the colorant (dye) in the ink. It must reside on the top of the paper in order to produce saturated colors. To achieve this, several vendors have developed special inkjet paper. These factors are the primary reasons why inkjet is not considered a plain-paper process.

One very distinct advantage of the liquid ink products is their ability to make very high quality color transparencies. The high quality is a result of using dyes as the colorants, instead of pigment particles that are used in thermal transfer. Pigment particles scatter light, and in a projected transparency this results in an image that is dark and much less colorful than its paper counterpart.

Continuous inkjet technology was pioneered in low-resolution direct mail printers. In an entirely different class is the Iris 3024, the first color configuration of continuous jet technology. Continuous inkjet systems operate by forcing pressurized ink in a cylinder through nozzles in a continuous stream.

The ink stream is unstable, breaking into individual droplets that either reach the page in the desired pattern or are deflected into a "gutter." Because each printed pixel is composed of multiple droplets, it is possible to control the pixel size, resulting in color halftoning capabilities such as those offered by Iris and Hitachi.

Permanence of images is an important issue facing inkjet technology. When speaking of permanence we generally mean:

- Resistance of image smearing when water is applied
- The ability to retain colors after exposure to light
- Image resistance to mechanical factors

Color images made from water-based inks usually are very susceptible to water spilled on them. Hot-melt inkjet does not have any waterfastness problems. Most inkjet inks use dyes as the colorant material — and dyes are known to fade. A crucial question is how long will the images retain their original color. Some inkjet systems have exhibited fading under office lighting within months, while others show only small color changes over a period of years. Careful formulation of water-based inks is leading to increased permanence for inkjet images.

Jet clogging, which necessitated complex and costly print head maintenance stations, has for the most part disappeared in the newer products. Solid inkjet seems to eliminate most of the reliability issues associated with inkjet. For the long run, inkjet is probably the strongest competitor to electrophotography for high-volume printing — largely because of its low print cost. Large print volumes will bring the cost of the special paper down, and inks, even the solid inks, are basically inexpensive.

**Thermal Transfer**

A color printing technology that has seen significant growth, in terms of units sold, is thermal transfer printing. Thermal printing has a varied history, but the technology has gravitated into two forms, categorized in terms of their imaging materials:
- Thermal wax transfer, in which the imaging material is a pigmented "wax"
- Thermal dye transfer, in which colored dyes comprise the imaging material

The basic components of a thermal transfer device are:
- A thermal print head comprised of a series of small heating elements spaced at 200 to 300 per inch
- Ink/dye donor sheet, the source of the image colorant
- Paper or receiver sheet
- A pressure roll

Applying an electrical current to the resistive heating element in the thermal print head causes it to become hot. The thermal head is in intimate contact with the donor sheet — assured by a pressure roll or platen opposite to the print head. As the imaging material heats up (about 100°C), it melts the wax. In the case of thermal dye transfer, the heat (300°C) drives a dye into the paper.

Distinct differences exist between the two methods of image formation. Wax transfer is a mechanical process; the softened ink must physically transfer and be attached to the paper. Thermal dye transfer, sometimes call dye diffusion, is a thermodynamic process where a color former (dye) is driven by the heat energy of the print head into the paper. A second, and more important, distinction is the amount of colorant transferred:
- Thermal wax transfer is bilevel in that only two levels can be produced; ink is transferred or no ink is transferred to the paper.
- Thermal dye transfer — a continuum of dye can be transported to the receiver or paper according to the amount of heat energy applied.

Image and print size for both methods are limited by print head size and the mechanics to move the ink donor web and paper. An 11-in. head seems to be about the maximum size available, although larger heads are possible.

Neither of these thermal transfer techniques can be considered "plain paper." The wax-based method requires smooth paper for complete transfer of the colorant to the paper. Rough papers inhibit the complete transfer of the dot, which results in a loss of print quality. With both thermal transfer methods, colors are formed by sequentially printing cyan, magenta, and yellow colorants on top of one another.

To achieve a broad range of colors, the wax-based method requires some halftoning, or dithering, since it is a bilevel process. Thus, resolution or dot addressability must be at least 300 dpi to produce high print quality.

Use of dyes rather than pigments is a major distinction between thermal dye transfer and other color printing technologies, notably conventional wax-based thermal transfer. In pigment-based systems, most light in the color spectrum is absorbed by the printed surface, and the light that is not absorbed (i.e., is reflected) is visible to the eye. Dye-based imaging systems allow the light to pass through the colorant, reflect off the printing substrate, and reflect back through the dyes (where unwanted colors are absorbed), resulting in more brilliant color tones.

## Thermal Wax

Thermal wax transfer is somewhat more costly than inkjet. The higher-quality thermal dye transfer, with its specially coated paper and multi-layer dye donor web, is significantly more expensive. Thermal transfer has low imaging material utilization. Only 30% to 60% of each of the colorant layers is transferred — the remainder is thrown away.

There have been some attempts to re-coat the ink donor sheet in the printer. Since thermal transfer is a multi-part consumable process, per-page costs are not likely to decline to the level of inkjet and electrophotography in the near future.

The mechanics of moving the ink donor web and paper has been developing over a number of years: it is a mature technology and therefore is very reliable. Perhaps the weakest link, from the reliability perspective, is the print head itself. If one of the heating elements should fail, a white streak is created in the print, and the print head must be replaced. The number of dots per heater element that can typically be printed before head failure is over 100 million.

## Thermal Dye

Thermal dye transfer is a continuous process. Because of its ability to transfer different amounts of colorant, according to the total heat energy applied, it does not require halftoning techniques. In fact if the number of levels is sufficient, say about 32 or more, a 200-dpi image can pass for "photographic quality."

The colors are placed sequentially along the ribbon length. As the ribbon is indexed past the print head, dye from each page-sized color block is transferred to the receiver. After the yellow for an image is transferred, magenta is overlaid on it, followed by cyan. Some ribbons incorporate black timing marks on the edge of the ribbon to ensure that each set of dots is precisely registered on the corresponding dots already printed in another color. Most thermal dye transfer ribbons are spooled from a supply reel to a take-up reel, often housed in cartridges for easy loading, straight feeding, and protection to the ribbon itself.

Thermal dye transfer is also called "dye diffusion" and "dye sublimation" in attempts to explain the actual process of migrating dye molecules from donor to receiver sheet. Thermal dye transfer technology shows promise in several emerging color applications, including video/electronic camera output, prepress proofing, and medical imaging.

It is unlikely to compete directly against solid inkjet technology in many environments because it shares very few design goals. The most critical distinction is that, whereas solid inkjet design is based on an assumption that the major market opportunities require true plain-paper printing, thermal dye transfer leverages off of the print quality benefits of specialized materials. Whereas dithering algorithms could permit solid inkjet products to produce various color shades (although with some reduction of effective resolution), thermal dye transfer produces near-continuous tones.

However, in addition to far higher supplies costs, there are considerable performance costs: an "A" size print takes up to 3 minutes to complete. The plain paper/higher speed vs. superior print quality issues may lead to interesting market battles in the few areas of application overlap between solid inkjet and thermal dye transfer.

Thermal dye transfer is a technology of interactions. It depends on the interaction of imaging materials with precisely controlled increments in a thermal print head. It depends on chemical interaction between the donor ribbon coatings and the receiver sheet coatings. These interactions are very sensitive and must be carefully controlled to ensure good print quality via exact transfer of dyes to create the desired color tones. Design of ribbon and receiver materials, and their associated dyes and other coating components, represents the most complex part of the development of thermal dye transfer technology. The companies with the wherewithal to develop the supplies will be the driving forces. The base materials and coatings of thermal dye transfer ribbons and/or receivers must have the following properties:

- Dyes that combine to create the desired range of color tones
- Carefully specified levels of reflectance to obtain desired optical density for the output
- Consistent response to print-head heat levels
- Ability to withstand print-head temperatures as high as 300°C (572°F) without distortion, curling, or darkening
- Resistance to static electricity during movement through the printing mechanism
- Lubrication to protect the print head from abrasion

In order to achieve consistent color tones and overall image stability, the dye donor and receiver sheets must be carefully matched. It is not possible to use the ribbon from one vendor coupled with the receiver from another. To some degree, matching should also take place between a printing system and supplies, since software-controlled print-head electronics can maximize print quality for particular materials; however, it is theoretically possible to use generic supplies with any printing system.

One of the biggest considerations in the choice of a dye set is the issue of image stability. Dyes must satisfy two opposed requirements. They must be mobile for printing, with molecules migrating from the donor to the receiver sheet in response to heat. Yet they must offer stability after printing when exposed to heat (which could prompt image retransfer), liquid (which could wash dyes out of the surface), or light (which could prompt fading). Kodak and others target an image life of about 18 years—roughly equivalent to that of conventional photographic prints, which use closely related imaging dyes.

Thermal dye transfer printing places tremendous demands on its receiving papers. To yield desired print quality while standing up to the rigors of printing and environmental factors, receiving materials, like donor materials, have multiple layers with strong adhesion between layers. Together the layers should provide the following characteristics:

- Controlled uptake of dye into the receiver coating at a rate that equates to a specified amount of heat energy, producing the desired color tones (more heat equates to more transferred dye, resulting in a darker color)
- High gloss, resulting in output that resembles a photograph
- Flat output with no relief image
- Tolerance of heat levels up to 300°C (572°F), and tolerance of considerable pressure when passing the materials under the thermal print head
- Ability to resist surface distortion and curling despite uneven application of heat
- Low sensitivity to light, dust, dirt, donor defects, abrasion, scratches, liquid, and other environmental factors
- Surface coatings that include solvents which prevent dyes from recrystallizing
- Image fixing techniques that prevent dye retransfer and are transient to the user; chemical fixing is preferable to a second heating or lamination step

There are typically three functional layers in a receiver sheet:
- The dye receiver layer, which accepts dye molecules
- The compliant layer, which is a cushion that permits even transfer of dye around surface imperfections such as dirt
- The support layer or base

The paper is generally very thick due to the coatings and the use of heavy base stocks to ensure flat printing despite heat and pressure. The dye receiver layer is at the top of the sheet, closest to the dye/binder layer of the ribbon. It consists of a coating that accepts the dyes while yielding the desired gloss characteristics. Coatings must halt the natural tendency of dye molecules to move, lending stability and thereby keeping printed dyes from running together.

Thermal dye transfer ribbons consist of multiple parts, typically a polyester base coated with several functional layers. It is the complexity of this structure, and choice of the chemical components in each layer, that makes the development and production of thermal dye transfer ribbons a specialized and difficult task.

## Summary

As a technology, electrophotography has the advantage in the high-volume color copying and printing areas. Two factors account for this: low unit-image cost and high process speed. It is unlikely that any of the existing color technologies will become competitive on these factors in the near future. As a consequence of high hardware costs, color electrophotographic technology will not migrate to lower volume/speed ranges in the near term.

The color hardcopy market has emerged much more slowly than many observers originally expected, plagued by limitations of output technologies, high costs, and the lack of a clear user mandate for graphic applications. Recently, acceptance of color has accelerated. Electrophotography will play a dual role in the development of the overall color market. Color copiers will promote use of all output technologies by facilitating the production of multiple copies for distribution. Color printers will offer plain-paper output at unprecedented speeds, making a scenario of color-based on-demand publishing realistic.

In the printing and publishing segment, the strongest position for color electrophotography will be for on-demand publishing, which is expected to grow as an extension of desktop publishing.

The color electrophotographic copiers and printers that are coming to market are part of an evolutionary process that dates back fifty years. Electrophotography has many significant advantages over competitive technologies: excellent output quality, high imaging and process speeds, low cost per page, and plain-paper capability. These attributes have extended monochrome electrophotographic configurations into many operating environments, with products ranging from sub-$1,000 personal copiers to 200+ page-per-minute laser printers. Developers of color systems hope to leverage the same attributes to gain a strong position in the color hardcopy arena, although the complexity of color may limit the product range.

Color electrophotography's development and market positioning is characterized by several factors:

- High R&D costs have limited participation to a few major imaging companies, such as Canon, Fuji, Xerox, and Japanese companies.
- Although price/performance gains have been made with each new generation of product, color electrophotography has a starting level of close to $20,000, which has restricted its market presence.

Monochrome electrophotography made its mark in the copier area, expanding into printing applications as reliability improved and software was developed that took advantage of the technology's strengths. A similar migration from copier to printer configurations happened with color electrophotography.

# Chapter 20

# Printing Technologies

The very heart and soul of the printing industry is the printing press. From its invention around 1450 by Johannes Gutenberg for letterpress printing and its metamorphosis to Senefelder's lithography, to the high-speed automated systems of today, we can think of no better way to introduce the printing process than to quote the International Paper *Pocket Pal*.

> "The production machines of the graphic arts industry are the printing presses. In general, a printing press must provide for secure and precise mounting of the image carrier (and, in lithography, a blanket); accurate positioning of the paper during printing; conveying the paper through the printing units to the delivery; storing and applying ink (and, in lithography, a dampening solution) to the plate; accurately setting printing pressures for transfer of the inked image to the paper; and means for feeding blank, or partially printed paper and for delivering printed paper."

Presses are either sheetfed or roll- (web-) fed. Much commercial work is printed on sheetfed presses. Magazines, newspapers books, and long-run commercial work are printed on web-fed presses.

Presses may be single color or multicolor. Usually each color on a multicolor press requires a separate complete printing unit of inking, plate, and impression mechanisms. A two-color press would have two such units, a four-color press would have four, etc. Some presses share a common impression mechanism among two or more printing units. Packaging and other special-purpose equipment may have combinations of lithographic, letterpress, and gravure units. A perfecting press is one that prints both sides of the paper in one pass through the press. Almost all web and many sheetfed presses are designed to perfect.

## Letterpress Printing

Letterpress is a relief process that uses a raised image. The raised surface is inked by rollers and then pressed against the paper to make the impression. In traditional letterpress, the text is printed from metal type and the pictures are printed from letterpress blocks. These elements are assembled together or imposed to create a form which sits inside a rigid frame called the chase.

It is one of the oldest forms of printing. It is not used much today. But if you walk around printing shops in some of the more remote areas, you can still find them used for business cards, stationery, labels, business forms, tickets, cartons, and especially newspapers.

## Duplicating

Originally a process using gelatin and solvent, stencil duplicating now uses a waxed stencil that is cut by a typewriter. It is also possible to produce photo stencils in which a picture or handwriting is photographed and a stencil made from that.

The process is used mainly in offices to produce circulars, forms, price lists and other items in which only a few hundred are needed and quality is not critical. Offset duplicators are true offset presses used mostly for short runs of black and white.

## Lithographic Printing

Lithography is the dominant printing process today. There are over 2,100 heatset web and almost 2,000 nonheatset web commercial presses in use in the United States. Almost 90% of heatset and about 80% of nonheatset web presses are multicolor, averaging over five units per press.

These numbers do not include web offset newspaper presses, many of which are printing newspaper inserts as well as newspapers. The growth in web offset has been slowed by the decline in periodical ad pages and some degree of overcapacity. After annual increases in growth of close to 15% in the decade up to 1990, the growth rate has declined to 7%.

Sheetfed offset lithographic printing is growing at an annual rate of about 2%. There are over 40,000 sheetfed presses, 17×22 in. or larger, 37% of which are multicolor with an average of 2.9 printing units per press in about 7,000 printing plants in the United States. In addition there are about 200,000 offset duplicating presses 21 in. or smaller in 45,000 commercial and quick printing plants, and 15,000 in-plant shops, averaging over 1.5 presses per plant.

About 5,000 offset duplicators were installed in 1994, but this has declined severely with the advent of copiers and electronic printers. Sheetfed offset's ability to

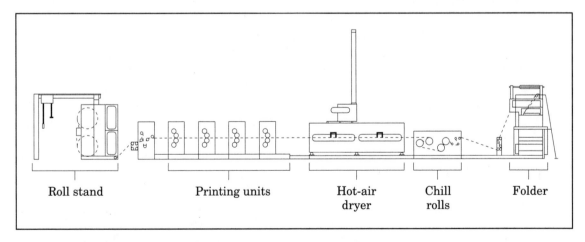

Roll stand          Printing units          Hot-air          Chill          Folder
                                             dryer            rolls

A four-color heatset web offset press.

produce short runs cost-effectively has kept them competitive with laser printers and copiers, but that advantage is fading as the newer devices integrate sorting and binding online.

The advent of on-demand and just-in-time concepts of production have led to the idea of producing the entire document. Although there have been automated duplicating systems, they have not been successful. The offset duplicator will continue to exist for the production of low-page-count or single-page documents.

Larger presses have been helped by developments like plate scanners and ink presetters, which speed makereadies, and in-line coating units on newer presses, which eliminate the need for antisetoff sprays.

Printing presses do not die. As mechanical devices, they are not obsoleted as often as electronic devices, and it is not uncommon to see presses that are 30 and 40 years old.

A major reason why offset presses will still play a role in the future will involve their adaptation to newer approaches, creating hybrid machines that combine offset principles but integrate image carrier or other reproduction technology. The GTO-DI is a perfect example. One can easily project that this technology could be adapted to larger presses. The AM ElectroBook press used lithography and electronic printing. Moore Research has developed technology that also combines traditional and digital printing. It is possible that printing presses of the future will routinely have inkjet or other printing technologies integrated for the printing of variable data.

The base of presses will be an important factor in the development of hybrid approaches. It is an identifiable and ready market for new approaches that reduce cost or provide new capabilities.

Lithographic presses are the most difficult to operate and control of the major printing processes. The ink/water balance is very difficult to monitor and control, and requires considerable skill and experience. Lithographic presses have more downtime, productivity is lower, and manning is higher than in flexography or gravure. Manning may consist of one person for each unit on the press.

New high-tech presses have lower manning, and unions have agreed to reducing manning on these presses, but even with these concessions, manning is generally higher than other printing processes.

Waterless offset has seen spotty growth in the last few years. It increases quality and removes one variable (ink/water balance) while adding another (temperature). There is no doubt that waterless will play a role in the future, but its percentage of use will be affected by increasing changes in water-based technology. Waterless can only grow if water-based technology stands still, and that is not likely.

## Flexographic Printing

Flexography is the youngest of the major printing processes. It is a relief process like letterpress except that it uses flexible, resilient plates and low-viscosity liquid inks dispensed by simple inking systems using engraved, or anilox, rollers. It started as the "rubber stamp" process (aniline printing) because the inks used were composed of soluble aniline dyes (which are toxic substances). The process requires a uniform plate, perfect cylinder, and uniform thick mounting tape to maintain the proper quality levels.

Even though flexography is a relief process, it is not plagued by makeready problems because it uses flexible, resilient plates that compress enough during printing to even out the impression. Interest in and use of flexography has spread from packaging to other printing markets. First there were paperback books followed by newspapers and phone directories.

Flexographic printing has many advantages, especially for products using water-base inks like newspapers, inserts, directories, and other paper products. Water-base inks have little or no rub-off so hands stay cleaner when reading newspapers, and they exhibit reduced environmental problems. They dry on the surface of the paper with no strike-through and very little, if any, show-through. This allows thinner papers to be used. Flexographic presses are simple to operate and require less manning than litho presses.

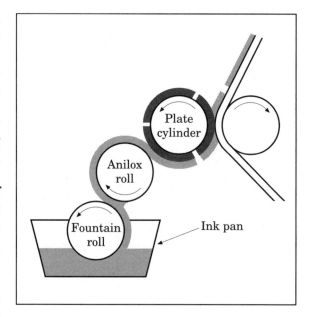

The basic principles of flexography.

Flexography also has some disadvantages. The relief plates used for flexographic printing are more expensive and slower and more difficult to make than lithographic plates. They do not reproduce as small a highlight dot or as smooth a gradation of tones throughout the tonal range as lithography.

Plugging and fill-in, which is also characteristic of letterpress printing, is still a problem though considerably reduced. Ink management is more complex than with offset, and more costly. Water-based inks, a major benefit of flexo, are also a drawback. They are caustic, corrosive, difficult to control, difficult to clean up, and difficult to de-ink. All these problems are under attack.

## Gravure Printing

Gravure has traditionally been a printing process for the production of long-run products in quantities beyond the capabilities of the other printing processes. The top 10 consumer magazines are printed by gravure — 15 gravure magazines account for 64% of the circulation of the top 25 magazines.

The 25 magazines printed entirely or in part by gravure represent 50% of all magazine circulation and a third of all magazine advertising revenues. Gravure accounts for 31% of the catalog market and is used for 75% of national inserts and one-third of all inserts. In packaging, gravure accounts for 29% of the receipts for folding cartons and 35% of flexible packaging.

Gravure has long dominated the long-run markets, but these markets have about reached their peak, and growth rates are now fairly static. A dominant market trend that has affected all printing markets, except subscription periodicals, mail order catalogs, and telephone and other wide distribution directories, is decreasing press run lengths due to demographic editions, target marketing, special promotions, and smaller inventories.

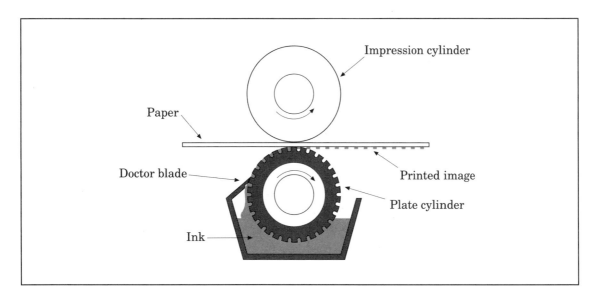

The basic principles of gravure printing.

Gravure in the U.S. is not geared to short-run publication, catalog, and directory printing. In Europe, runs of 150,000 by gravure are commonplace; in the U.S., gravure printers cannot break the half-million barrier profitably. There are a number of critical issues in gravure that relate to its inability to compete with web offset in the shorter-run publication marketplace: overcapacity due to the addition of new, wider, and faster presses without retiring older, inefficient presses they replace is one issue, and press makeready downtime and delay time is another.

Another serious limitation to expansive growth is high investment costs for gravure presses and installations due to the use of solvent inks that create pollution problems and fire and explosion hazards. These limitations could be eliminated by the development of satisfactory water-based or radiation-curing inks. If work going on in these areas is successful, gravure could begin to command a larger share of the printing market.

Gravure has a number of significant advantages. It is a simpler process to control than lithography, and it is easier to automate. It is the first process to use computer-to-cylinder technology. Press automation is advancing.

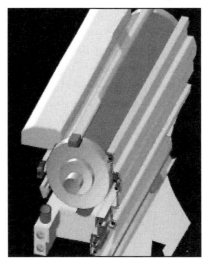

*Courtesy of Agfa, a division of Bayer USA*

## Digital Printing

In 1992 Heidelberg released the GTO-DI, the first printing system to introduce the concept of on-demand color reproduction. In concept, it created printing plates on a traditional printing press, effectively solving the registration problem associated with the four-color printing process. The plates were created with a spark discharge that then allowed them to print without water, effectively solving the ink and water balance problems associated with offset lithography.

Today the digital color press manufacturers include Indigo of Israel and Agfa/AM/Xeikon of Belgium. Although controversial, some experts are beginning to include color copiers, such as the Canon CLCs or Xerox Majestiks, in the digital press category. The digital black-and-white products include the Kodak Lionheart and Xerox DocuTech.

# Chapter 21

# Changing Markets for Print

In 1989, Mike Bruno, the well-known dean of graphic arts consultants, attempted to predict the use of different printing processes in the future. He essentially saw gravure, flexography, and lithography each with a 25% market share and the balance divided among other reproduction methods. We have a different perspective. In this chapter, we try to fearlessly predict the future.

Printing traditionally involves pressure (impact) processes using plates or some image carrier, imaging material (usually ink), and pressure between the image carrier and the substrate, usually paper, to reproduce the image.

There are also pressureless, or nonimpact, processes, like electronic, electrostatic, magnetographic, ion deposition, and inkjet printing. We feel confident that reproduction of material on substrates for defined audiences will still be around in 2004. Each reproduction technology will find a niche, where it offers the price and performance characteristics that are required.

## Market Projections

The new prepress technologies have application to all the processes. Lithography and gravure are sharing about equally in their benefits. New press controls and long-run and high-exposure-speed plates are favoring lithography for publication markets. Management information systems are increasing productivity and reducing waste and cost of lithographic printing. New developments in flexography and gravure could help them challenge lithography's lead after 1995 and into the next century if these processes can overcome some of their limitations.

It is predicted that the communications industry dedicated to disseminating images of text and graphics on a paper-like substrate will be divided into these parts: (1) lithographic-like process, (2) gravure-like process, (3) flexographic-like process, (4) letterpress, screen, and other plate-based processes, (5) electronic, inkjet, and other new plateless printing processes, and (6) hybrid processes.

It is almost impossible to predict that any one technology for reproduction will "win" since all of them are improving their quality, capability, and cost-effectiveness.

## Projection of Changes in the Number of Colors

It is predicted that the easier it is to reproduce color, the more that users will demand color:

|  | 1994 | 2004 |
|---|---|---|
| B&W | 56% | 45% |
| B +1C | 10% | 7% |
| 4C | 16% | 35% |
| 6C | 14% | 9% |
| 8C or more | 4% | 4% |
| Total | 100% | 100% |

## Changes in the Type of Printing

We do not believe that any of the existing reproduction technologies will disappear. Rather, users will be able to select from a wide variety of approaches:

|  | 1994 | 2004 |
|---|---|---|
| Offset | 55% | 35% |
| Printer | 10% | 25% |
| Copier | 15% | 20% |
| Gravure | 10% | 10% |
| Flexo | 5% | 5% |
| Other | 5% | 5% |
| Total | 100% | 100% |

## Projection of Type of Printing

Here is a fearless projection of these technologies. Note that all will be able to work with digital files and that most are applying new technology to make the process more efficient and cost-effective:

|  | New technology being applied | Works with digital prepress | Long-range forecast |
|---|---|---|---|
| Lithography |  |  |  |
| Water-based | Yes | Yes | Some loss |
| Waterless | Yes | Yes | Some growth |
| Gravure | Some | Yes | Static |
| Flexography | Some | Yes | Static |
| Letterpress | No | No | Loss |
| Screen printing | Yes | Yes | Static |
| Other plate processes | Yes | Probably | Unknown |
| Electronic | Yes | Yes | Growth |
| Inkjet | Yes | Yes | Growth |
| Other plateless processes | Yes | Yes | Growth |
| Hybrid processes | Some | Yes | Growth |

## Distribution by Process — Printing, Publishing, and Packaging

|  | 1989 | 1995 | 2000 | 2025 |
|---|---|---|---|---|
| Offset lithography | 47% | 47% | 45% | 30% |
| Water-based | 47% | 46% | 37% | 20% |
| Waterless | 0% | 1% | 8% | 10% |
| Gravure | 19% | 20% | 20% | 20% |
| Flexography | 17% | 18% | 19% | 20% |
| Letterpress | 11% | 7% | 4% | 2% |
| Screen printing and other plate processes | 3% | 3% | 3% | 2% |
| Electronic, inkjet, and plateless processes | 3% | 5% | 9% | 26% |
| Total | 100% | 100% | 100% | 100% |

*Source: GAMA*

## Moving Toward Shorter Runs

Some of these long-run market areas may also have short-run requirements. The trend toward elimination of inventories and printing shorter runs has created many new markets for short-run printed products.

It is important to note that there are market niches based on run length and resolution — and reproduction technology. The chart below shows nine reproduction technologies, most in color. There are certain areas where each has a production and economic advantage. There are also an increasing number of areas where they overlap. This is indicative of the new world order that is evolving as new and existing technologies adapt to the short-run market demands.

The two axes are simply price up and down, and capability left to right. As we go up and to the right the systems get more expensive but also more capable.

Personal, desktop, and network color printers are all related. The personal devices are mostly inkjet or bubble jet technology and priced well under $1,000. Their speed is rated in minutes per page rather than pages per minute. Desktop devices provide more speed, about 2–5 pages per minute and tend to be electrophotographic, although there are some using dry ink. Network versions have two paper trays that are standard and speeds that go beyond 5 pages per minute.

Color copiers are included because they define a distinct class of reproduction systems. You can add a RIP to a color copier and its becomes a *de facto* color printer, or

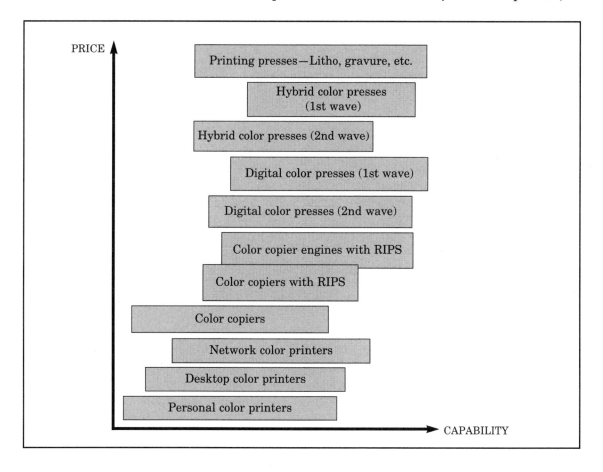

you can remove the color scanner and have a "pure" color printer. This last category will see systems in the 20–30 page-per-minute range.

Digital color presses are defined by their speed and quality. Indigo and Agfa/AM/Xeikon are in this area. They automatically duplex and provide enhanced finishing features. The first wave refers to systems in the $350,000–450,000 range, which will continue to move up-market — that is, increase in capability. We expect a second wave that will bring the technology down to $200,000–300,000.

Hybrid color presses are printing presses with some form of on-press platemaking. The first wave would refer to the $700,000 GTO-DI and versions that move to larger sheet sizes. The second wave refers to the Heidelberg Quickmaster-DI, which brings the technology somewhat down-market. Lastly there are traditional color printing systems.

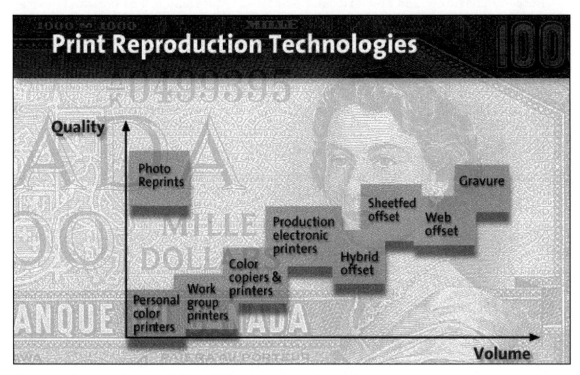

*Courtesy of Agfa, a division of Bayer USA*

# Section IV

# Specific Devices

# Chapter 22

---

# T/R Systems MicroPress

---

When T/R Systems unveiled its first MicroPress system at the 1995 Fall Seybold show in San Francisco, it was an unknown. At the time, the four-year-old company was better known in laser printer engineering circles or within Colorocs, a color copier company where the founders previously worked.

T/R Systems was founded in 1991 by Neal Tompkins and Frank Rowe, both former Colorocs staff. Tompkins was vice president of development, and Rowe was an executive vice president. T/R Systems started as an engineering service company, providing technical assistance to well-known laser printer suppliers.

In 1993, with a few years of success and their own engineering prowess, the founders decided to create their own product and obtained $6 million in venture capital. In April 1995, Rowe recruited Mike Daly as vice president of sales and marketing. With experience at Compugraphic (now Agfa), Varityper (now Prepress Direct), and Barco, Daly was the last piece in the T/R Systems senior staff triumvirate. Shortly after Daly joined, the MicroPress 312 was introduced at the 1995 Fall Seybold Show, and the MicroPress 24 was introduced at the On-Demand show in April 1996.

## Overview

Similar to many other products listed in this book, the MicroPress products are unusual, hybrid technologies that do not fit into a predefined category. Other unusual, hybrid technologies discussed in other chapters are the presses equipped with Presstek technology as well as the hybrid technologies contained within the Indigo E-Print.

The uncommon aspect of the T/R Systems MicroPress is that it uses traditional laser copy machine print engines (i.e., Canon P320 and P550) but with an untraditional distribution of one raster image processor (RIP) with several writing engines. In contrast, typical copiers and printers use one RIP with one writing engine. As discussed at the end of this chapter, the result of this unusual configuration is a lower and more attractive price/performance ratio.

The PressDirector uses a 133-MHz Pentium-based PC with 128 MB of RAM and a 4-GB disk, running the Windows NT operating system. This is now a popular configuration because it allows the multitasking of Unix without the cost of a Sun workstation. This is important since it allows the RIP to be ripping one job while outputting another. The RIP software is from Harlequin. You rip the file once and the software then runs the selected number of devices, automatically collating sets if necessary and if specified. The software also assists the operator in "work-and-turn" duplexed printing since the pages must be manually turned over and reloaded.

Frank Romano refers to this system as the device that works like Christmas tree lights. The system is comprised of either two or four Canon 320 color printer engines linked to a common RIP/server. Apple and Tektronix use this printer for their $7,000 desktop color offerings. EFI sells a higher-priced package.

At 3 standard color pages per minute, the printer will not set the world on fire, but a $57,000 package that provides 12 pages per minute with an estimated $0.50 cost per copy may create a "poor man's color press." The cost per copy is based on our continuing refinement of a model that takes all capital, operating, consumables, and human costs into consideration.The resolution is 600×600 dpi with 32-bit color. Each printer is rated at 40,000 copies per month.

The two-printer system comes in at $0.49 per page. As we are discovering, the capital cost is less important than the speed of the system and the cost of consumables. By using a monocomponent toner, Canon has made the system much more efficient. A lot of the printing technology seems to have come from the CLC, fuser oil in particular. Like most Canon devices, major elements are replaced, which essentially provides a minor overhaul.

The following table lists the costs of the various consumables. Apple's prices for these items are slightly higher.

| Toner | | |
|---|---|---|
| Black | $ 95 | |
| Y, M, C | $110 each | 4,000 impressions at 5% coverage |
| Fuser oil | $ 21 | 4,000 impressions |
| Fuser assembly | $576 | 60,000 impressions |
| Photoconductor | $126 | 40,000 impressions |

As mentioned previously, the speed is faster for black-only pages: 12 pages per minute versus 3 color pages. Actually the higher speed is available for any page with only one of the toners on it — it does not have to be black.

A sheet of paper is moved to a transfer cylinder where the leading edge is gripped by plastic fingers. The trailing edge is held down by electrostatic forces. The image is then written to a photoconductor drum by laser. After the image is written as a set of negative charges, the toner rollers come in contact with the photoconductor and toner is attracted to the charged image. The photoconductor continues its rotation, and the toned image is transferred to the paper.

The circumference of the photoconductor is about half the area required for a letter-size page, so the entire process is repeated twice for each color. Thus, there are eight rotations of the photoconductor drum for each full-color page. Now you see why the speed is 3 color pages and 12 monochrome pages per minute. After all the toners have been transferred, the sheet is released from the transfer cylinder and sent through the fuser and out.

## Print Engines

The print engine in the MicroPress 312 is the P320 color engine from Canon, the same engine used in the Apple Color LaserWriter and the Lexmark Optra C printer. It has a print speed of 12 ppm in black-and-white mode, or 3 ppm in four-color mode.

The maximum printing resolution is 600×600 dpi, with 8 bits per pixel/color, although the RIP has the capability of rasterizing data at 1200 dpi, downsampling to 600 dpi, and antialiasing text and line work for smoother appearance.

The input media options include paper ranging from 16- to 28-lb. paper and transparency material. Paper sizes include 8.5×11-in., 8.5×14-in., and A4 cut-sheet paper sizes. There are three input trays: a 250-sheet input cassette, 100-sheet manual feed slot, and a optional 250-sheet tray.

The engine in the MicroPress 024 is the same 600-dpi engine used in the Hewlett-Packard LaserJet 5Si MX. Each engine prints on 11×17-in. sheets in the two-sided (duplex) mode at a rate of 24 (letter-sized) pages per minute.

The system can come with a minimum of two engines and a maximum of six using the current Pentium controller. According to T/R Systems, the Pentium Pro controller scheduled for release in the fourth quarter of 1996 will support up to eight print engines.

The six-engine configuration prints 144 ppm, which is slightly faster than the DocuTech 135. According to T/R systems, they are not trying to compete with the DocuTech because the DocuTech has specialized features such as job management, finishing, cover insertion, folding, and stitching not available on the MicroPress.

## PressDirector

Another way to describe this unique ratio of RIPs to writing engines is as a parallel architecture or, as T/R systems calls it, a "Cluster Printing System." According to T/R Systems, its Cluster Printing System is made of one RIP/server station called the PressDirector and multiple print engines called PrintStations.

The PressDirector is responsible for communication and monitoring of the print engines, print spooling, OPI (Open Prepress Interface) functions, ripping, and the overall management of the system. The PrintDirector software runs on a Windows NT platform and accepts files from the following platforms: Microsoft Windows 3.1 or later, Windows 95, Windows NT 3.1 or later, and Apple Macintosh 7.0 or later.

The PressDirector uses a multitasking software package running on Microsoft's NT Advanced Server software platform. It runs on a 133-MHz Pentium computer with 128 MB of RAM, a 4-GB hard drive, and a CD-ROM drive. After initially offering a 64-MB configuration, T/R Systems found that 128 MB of RAM was required to print files of any complexity to any engine configuration.

The 64 MB of RAM is enough for moderate numbers of pages of moderate complexity. We estimate that moderate numbers of pages is 20 pages /day and that moderate complexity is less than an average of 10 MB of data. For more pages or larger pages, T/R Systems recommends upgrading to 128 MB of RAM.

Within each of the print engines are control boards designed by T/R Systems and interfaced through the PC's PCI bus. Each board can control up to two engines. These controller boards use real-time compression that can achieve 120-MB/sec. transfer rates to the PrintStations. Besides some additional hardware added by T/R Systems, the writing engines only have writing engine and network input ability. This results in faster network transmission and decreased storage demands.

## Multiple Runs

In situations in which the same job will be printed several times, users have the option of storing the file on the disk, as space allows. The ripped files are stored in a compressed bitmap form and the decompression occurs on the fly in the print engines. After rasterization, the files are sent to the job parser and printing manager software (PressDirector), which distributes the job to the various print engines and monitors the status for any printing problems (i.e., out of paper). Once successfully imaged, the job is removed from the printing queue and the next job is sent.

One feature particularly important to the MicroPress products is the slip sheet feature that inserts slip sheets between jobs. Although useful in many copying environments to separate the pages sitting in the output bin, it is particularly important for the MicroPress products because one job may be split across the different output stacks on different output engines. Therefore, the user may need to "thumb-through" the output piles and distinguish the appropriate jobs.

In addition, the system can be set up to print job information and well as a color band across the bottom of each slip sheet. For the ultimate separation of jobs, oversized sheets (i.e., A4) can be inserted through the manual feeder.

## OPI

The PressDirector software also offers OPI (Open Prepress Interface) functionality to prevent network overloads. The problem is that if several users send large color documents to the printer it can grind a normal network to an agonizing halt. To combat this problem, spoolers, queues, and OPI are used.

Developed in 1989 by Aldus Corp. (now Adobe), OPI is a set of PostScript specifications for use in desktop publishing, networked printing, and color separation software. It substitutes a small, low-resolution version of a graphics file for a high-resolution original until just before a document reaches its output destination. PressDirector supports OPI with both TIFF and EPS file formats.

Using OPI can improve productivity. There is a direct relationship between a file's resolution (dpi), size, and output time. As the resolution of the file increases (dpi) so too does the size (MB) as well as the output time (min.). This becomes a problem in the early stages of proofing because you might only need to see low-resolution versions.

In some OPI strategies, low-resolution files are placed within the document and, then depending on which queue is used, either a low- or high-resolution file is output. T/R Systems has implemented this solution. The PressDirector software allows the operator to configure different queues with different ripping resolutions. Therefore a first proof could theoretically be printed to a queue that ripped the files for 300 dpi, 6 bits/pixel output for faster processing while the final page could be printed at 600 dpi with 8 bits/pixel output for higher quality.

Two preset queues are a 600- and 1200-dpi queue. When outputting four-color jobs with text for final output, the 1200-dpi queue causes the system to rip the file at 1200 dpi and then downsample the file to 600 dpi using antialiasing techniques to smooth out the text. If the file has no text, the 600-dpi queue would be appropriate.

There is also a soft proofing option. Files that have been ripped can be viewed on the screen at the PressDirector. Users can look at various levels of magnification to inspect jobs before printing.

## Color Consistency

While the "one-RIP, multiple-engine output" strategy has advantages in the price/performance ratio (discussed at end), the greatest potential disadvantage is color consistency. If you print out 20 pages from each of the four writing engines, will they match? There are two sides to this argument as well as the T/R Systems mechanisms to address this issue.

The advantage of using multiple engines is that there should be less "drift" in the color within each engine because the run lengths are less. In contrast, if you printed 80 pages (20 pages × 4 engines) from one device, the toner would become depleted and the color would drift more when comparing page 1 to page 80.

The disadvantage of the multiple engine strategy is that each engine has to print the same colors similarly. Let's say that these 80 pages are 20 four-page newsletters. The fastest way to print this would be to print page one on device one, page two on device two, etc. If each page had the same color bar on the page, they would all have to match.

To address this issue, the MicroPress prints and monitors two continuous-tone control bars on the nonprinting area of the drum of each print engine. Also included with the system is a separate color control test form. Users can print out the form, read the printed values on a strip-reading densitometer (i.e., X-Rite), and load the values into a correction table. Separate gamma compensation curves are maintained for each print engine with the Print Director software and corrections are applied as required to maintain consistent results.

## User Experience

This unique ratio of RIPs to writing engines does not change the way the system looks to the user or user's experience. The jobs are sent or "printed" to the MicroPress over the network. The output jobs are then processed by a print spooler and placed into a print queue.

Once the job is placed in the printing queue, the user's computer screen is freed up allowing the user to continue with other work. The user experience is the same regardless of whether the job was sent to a stand-alone copier/printer or to the Micro-Press. In fact for users accustomed to waiting for the local spooling of print jobs on their own computers, this is often faster.

Once the job is spooled, the user or shop manager can rearrange the printing order as priorities change. After each job is ripped, it is sent to the printing engines and stored in RAM and/or the hard disk depending on the amount of RAM and the file size. The only difference between the MicroPress and several stand-alone copier/printers is that the one-RIP, one-engine devices could each rip a file while the MicroPress only rips one at a time.

The PressDirector interface is a well-designed graphical user interface (GUI) using icons and a consistent look and feel. On-line help is available and is organized in a hierarchical menu strategy allowing users to progressively access more detailed information. Similar to the icon-based approach on the front panel of many copy machines today, the interface simplifies error reporting. For example, the "paper jam" display shows where the paper jam occurred.

The PressDirector software keeps track of job names, printing parameters per job, number of pages, print status, and number of copies made. The PressDirector also keeps in-depth records useful for maintenance, utilization calculations, billing, and consumables replenishment. For example, information about the toner level, number of pages (from each current photoconductor drum), and the total number of pages per printing engine is available. In addition, the system can report the amount of toner used per job and ripping time. Users could charge different rates based on the toner coverage and ripping time.

Like other print queues, the spooler allows users to change the order of files and manage previously ripped files. A feature typically associated with more-expensive copier/printers is the job scripting feature that allows the operator to selectively assemble elements from different jobs into a new document. Another useful feature is the ability for Macintosh users on the network to access a status window and see the status within the print queue.

Another advanced feature is the ability to print variable data. These devices take advantage of the PostScript Level 2 "form feature" capability within the Harlequin RIP, which allows a background form and a separate variable data stream to be merged in the ripping process. Users achieve variable printing by building an output form with an input data stream of variable data.

### Service

If you already have a color copier service or work with anyone who sells this service, you may have discussed the issue of duty cycles (rated pages/month) or service (repair staff visits). For companies using equipment with small duty cycles, high volumes of output, or service contracts, it is not unusual to be on a first-name basis with the service technician. For some companies, the expense in service warrants hiring a full-time staff member.

The print engines in the MicroPress 312, the Canon P320, are designed as a low-cost, user-serviceable design. A fairly unique concept is that the photoconductor drum is separate from the toner cartridges (allowing individual replacement), each toner color is a separate cartridge, and the fusing section can be replaced in 15 minutes with a simple service procedure.

In addition, the writing engine is essentially a disposable unit once the rated engine life is reached. According to T/R Systems, the useful engine life is about 250,000 prints per unit (1 million prints per four-PrintStation system). In comparison, Apple claims that its Color LaserWriter is good for 300,000 black-and-white or 150,000 color pages.

But as they say in the car commercials, "your mileage may vary depending on your driving." Just like a car driven on highways will last longer between maintenance, a copier run continuously for longer periods will last longer. Some experts say that, when the engine is run continuously rather than start-stop, it will at least double its useful life and prolong the photoconductor drum life too.

## Cost of Consumables

Cost is one of the most important as well as confusing aspects of on-demand and digital printing. Throughout the book we have tried to describe the costs, particularly the cost of consumables. For the T/R Systems color device (the MicroPress 312), the price of black toner cartridges is $94, and $100 for color cartridges. Each cartridge lasts for 4,000 impressions at 5% coverage, 2,000 at 10% coverage, and 1,000 at 20% coverage.

The photoconductor lasts for 40,000 impressions and costs $126. The fuser costs $565 and lasts for approximately 60,000 impressions, and the fuser oil costs $26 for 10,000 copies.

According to T/R Systems, the consumables cost per page is:
- 3% black page (typical all text page)    $0.027
- 5% black                                 $0.038
- 5% C,M,Y,K/ea. (20% coverage)            $0.134
- 20% C,M,Y,K/ea. (80% coverage)           $0.37

## The Latest — The RipStation, the SatellitePress, WebPress Software Suite

According to T/R Systems, three new products will be introduced at the 1996 Fall Seybold show in San Francisco: the RipStation, the SatellitePress, and the WebPress Software Suite.

One of the most common bottlenecks we see across all business is ripping time. Although better than a few years ago when files would run for hours, very large files can still take a long time. The RipStation is one of the first copier-based RIPs to use multiple RipStations to work on one job in parallel or multiple jobs simultaneously. The output of the multiple RipStations can be merged into one single job and printed on any full MicroPress or SatellitePress.

The SatellitePress represents a complementary product to the RipStation. It will accept data from any RipStation Manager and drive multiple PrintStations as one virtual printer. It can accept and merge compressed ripped print jobs, which can decrease the ripping time.

The WebPress Software allows the RipStation and the SatellitePress to work over the Internet or intranet. Each MicroPress or RipStation includes a web server and browser and WebPress Publishing Suite software.

## Summary

The unique MicroPress devices have some unique advantages and disadvantages. The disadvantages include short duty cycle, small paper size, manual duplexing, no on-line finishing, no sorter/stack, and no copier functions.

The 312 engines last only 250,000 impressions/engine, and the 024 engines last for 100,000 impressions/engine. In contrast, a DocuTech engine could last a million impressions. The engines in the 312 are limited to letter- or legal-sized pages and a simplex mode. More productive, the 024 can print on 11×17-in. paper and can print two-sided (duplex mode) pages. Neither machine offers on-line finishing, sorter, or stack options or the ability to make copies.

The greatest advantage of the MicroPress systems is the price/performance ratio. For example, with the MicroPress 312 each writing engine is capable of 3 ppm (page per minute) in color or 12 ppm in black-and-white. As illustrated below, the number of pages printed depends on the number of printing engines and whether the job is black-and-white or color.

| Number of PrintStations | B&W Speed (ppm) | Color Speed (ppm) |
| --- | --- | --- |
| 4 | 48 | 12 |
| 3 | 36 | 9 |
| 2 | 24 | 6 |

A four-PrintStation MicroPress lists for approximately $50,000, which is comparable to a stand-alone color copier or high-speed black-and-white printer but much less expensive than a digital press such as a $250,000–400,000 Indigo, Spontane, DocuColor, or Xeikon-based device.

Other advantages include modularity, flexible configurations, easy maintenance, low-cost consumables, and low downtime. The unique cluster printing configuration is modular allowing users to add print engines until the controller limit is reached. The unique cluster printing can be configured to allow certain users to print to certain engines or all users to print to all engines.

Perhaps the second best advantage of the MicroPress is the low cost to run. This is the result of simple maintenance and low-cost consumables. Almost across the board, other devices require a maintenance contract. For products struggling to differentiate themselves (i.e., Xeikon-based devices), the maintenance contract is one way of differentiating themselves. (See IBM chapter on "click" charges.) The Micro-Press, in contrast, is serviced by the user. It does not require a technician.

Other factor related to the low operating cost is the commercial availability of the toner and paper. Many of the other devices require the purchase of toner and paper from the manufacturer, which in many cases is more expense than commercially available products.

The MicroPress is a simple idea whose time may have come. Although it is limited to letter-size pages with a majority being simplex rather than duplex because of the productivity issues, there is probably a market with very small quick printers and others for this capability.

The two- or four-printer approach provides a form of backup for printing services. The systems will be sold by #1 Network, a consortium of graphic arts dealers who have banded together to handle advanced products. We are told that T/R Systems has sold out its production to the end of 1996. They are off to a good start.

# Chapter 23

# Xerox DocuTech

The DocuTech Publishing Series is a group of 600×600-dpi output products: the DocuTech Model 135, Model 90, and Model 6135. There are two basic engines: the Model 135 and 6135 operate at 135 pages per minute (ppm), and the Model 90 operates at 92 pages per minute. There are several differences between the three models. The Model 135 and 90 are designed as total production machines, with the ability to scan hard copy originals, accept electronic files, and perform cut-and-paste operations via a WYSIWYG user interface monitor. The Model 6135 accepts electronic files and is designed to operate as a high-speed printer, with the formatting of pages being performed upstream and then sent to the printer.

Maximum paper size of this series varies, with the Model 90 accepting paper up to 8.5×14 in. (216×356 mm), the Model 135 accepting paper up to 11×17 in. (279×432 mm), and the 6135 accepting paper up to 14.3×17 in. (363×432 mm).

Along with the different maximum size sheet, the print speed for the different models also varies. The Model 135 and 6135 can image an 8.5×11-in. (216×279-mm) simplex page at 135 pages per minute, while the Model 90 images the page at 92 pages per minute.

The Model 90 and 135 incorporate a gray scale reflective scanner, which scans at 600×400 dpi (eight bits deep, or 256 levels of gray), interpolated through software to 600×600 dpi, and which is capable of scanning hard copy line or photographic originals. The scanner can accommodate a maximum original size of 14.3×17 in. (363×432 mm). The operator can manually place originals on the glass or can use a document feeder that operates at 23 pages per minute. The scanner is capable of variable imaging size, from 10% to 200% in 1% increments. The scanner can be operated through the operator interface or via control buttons on the scanner itself. Because the DocuTech 90 and 135 printers are digital printers, the original need only be scanned once and the information is stored on the machine for printing as many copies as needed. The DocuTech 6135 does not have a built-in scanner and, as such, accepts digital files only. An optional, stand-alone system, Xerox Documents on Demand (XDOD), can be used for scanning and document management capabilities.

A second differentiating feature is methods of processing digital files. All three models can accept industry standard PostScript Level 2, PCL5, or TIFF formats. The DocuTech Models 90 and 135 can accept print-ready files in a non-network environment via a Media Server. The Media Server is an IBM 486 PC that accepts 3.5- or

Xerox DocuTech Production Publisher.
*Courtesy Xerox Corporation*

5.25-in. (89- or 133-mm) IBM-formatted floppy disks that contain the customer's print-ready files. This is a good solution if a lot of the files created are in a walk-up type of environment. All models of the DocuTech can also be in a networked environment. Networks supported include Novell, TCP/IP, and EtherTalk. A customer can "print" to these machines via a network as if his or her computer was physically attached to the printer. This gives the customer the ability to proof the job locally.

The raster image processor (RIP) is a major difference between the three models. The Model 6135 has a very fast RIP that is designed for customized, personalized types of documents and that is based on a Sun Sparc workstation. After rasterizing the pages, the files are transported to the 6135 printer via a video link cable.

The Model 90 and 135 uses either a RIP running on an IBM 486 PC, or a Sun Sparc workstation. The IBM 486 PC version has a typical speed of 6–8 pages per minute for image-intensive jobs and 45 ppm for pages of text. For an average page with images and text, the rate is approximately 24 pages per minute. The proprietary RIP board contains a 33-MHz Intel 80960CA RISC CPU, has 16 MB of RAM, and translates PCL 5 (supplied by Peerless) and PostScript Level 2 (supplied by Adobe). In addition, jobs formatted in Xerox's own Interpress language are passed directly to an Interpress interpreter on the DocuTech itself. The rasterized pages are moved to the printer via Xerox XNS network protocol.

The Network Server Plus operates on a Sun Sparc workstation Model 40, 85, or 61, which can be configured in three different power and memory levels. This station would be used in a high-page-count environment.

The Model 6135 is designed to have all of the page formatting performed before the rasterizing takes place. Hard copy originals can be scanned off-line using Xerox Documents on Demand (XDOD) hardware and software. After scanning and pagination, the resulting file would be sent to the 6135. The file that is printed on the 6135 usually contains variable data.

The Model 90 and Model 135 have a 17-in. (432-mm) black-and-white WYSIWYG user interface built into the actual machine. The operator communicates with the DocuTech via this interface, building a job ticket, performing cut-and-paste operations, and responding to messages about the current status of the machine. The interface is controlled via a touch-sensitive screen, a mouse, or keyboard. User friendly icons and pull-down menus, similar to conventional Macintosh or Windows computer screens, are used throughout. The operator begins a print job by filling out a job ticket. A job ticket contains information about the formatting of a print job. Once the job ticket is filled out, it can be stored for later retrieval. A job submitted via a network can have almost all of the information filled out by the customer, and this information can be modified by the operator. In the case of a hard copy job that will be scanned, all of the information must be indicated by the operator because the job is new. Typically the operator steps through the different areas of the job ticket screen one by one until the entire screen is filled out. The job ticket contains the job name, which can be changed from the name given by the submitter of a network job. The operator can choose to set up accounting numbers that can be used for billing purposes. However, if a customer submits a network job with an inaccurate accounting number, the job will refuse to print. The quantity, which can be up to 9,999, is selected next. Finishing considerations are addressed next. The finishing choices include deciding whether to have a multipage job either uncollated or collated and deciding whether a slip separator sheet should be inserted between sets. Collated sets can have a single or dual stitch applied. Tape binding can also be specified. If a SBM (Signature Booklet Maker) is attached to the DocuTech 135, the operator specifies that the job is going to be saddle-stitched. Page numbering can be turned on, and a location can be specified for the numbers. This is a very useful function for proofing purposes.

The next group of specifications refer to how the pages will appear and also what size paper will be used. If the scanner is being used, the type of original can be specified. Such considerations as a colored background or light text can be indicated. The choice can also be made if the original is text-only, text and photo, halftone, or text and halftone. By pointing out to the machine what type of original will be scanned, adjustments can be made in image quality during the scanning process. Because the scanner is a gray scale scanner, adjustments can also be made to the quality of the resulting halftone. Corrections can be made for overly dark or light original photographs. Sharpness can also be adjusted. The default setting results in a halftone that is 106 lines/in. (4.2 lines/mm). This can be changed to 85 or 65 lines/in. (3.3 or 2.6 lines/mm), and a straight-line screen can be also be specified. The next part of the job ticket concerns the paper that the job will be printed on. Size and color can be specified for the entire job, or on an individual-page level. Because the DocuTech has three paper drawers, up to three different papers can be specified. If more than three different papers are required, the operator can choose to either collate different papers together and then run them as sets automatically, or collate the job together after all the printing is completed.

Two-sided originals can be run as either two-sided or single-sided copies. Likewise single-sided originals can be turned into two-sided copies. A job may be rotated. The image or part of it can be shifted, which is important for some bindery considerations. If the job will be saddle-stitched, the individual pages will be automatically rearranged, shifted if necessary, and blank pages will be added to properly produce a saddle-stitched booklet.

Possibly the most useful aspect of the operator interface is the ability to perform cut-and-paste operations. Photographs can be merged with text, and blocks of text or other images can be copied from a page and then relocated to another page. This gives the user of a DocuTech the ability to perform last-minute changes to a document, or to reuse parts of one job in another job. A very popular use of this ability is if a customer does not own or doesn't know how to scan photographs. The customer will create his or her job via desktop publishing software and supply the page geometry with the text and graphics in place but will leave windows for the photos, similar to conventional camera work. The DocuTech operator will rip the job and then scan in the photographs, placing and cropping them with the DocuTech. The process of cut-and-paste can be performed either globally throughout the job, on specific pages, or down to one page. A logo, for example, may be used as a common element throughout a job. Electronic masks can also be applied to a job to eliminate unwanted blemishes or images. This is very useful with poor-quality originals.

The system allows you to build a default job ticket and to save up to 625 job tickets. This feature is also available in a networked situation, where a customer can build a job ticket and retrieve it later.

## Printing

The DocuTech 135 and 6135 print engine uses the engine from the Xerox 5090 duplicator, but it has a new paper path that enables duplex printing on 11×17-in. (279×432-mm) sheets.

The engine uses a revolving electrophotographic belt whose surface is charged as it revolves. The belt is large enough to hold seven 8.5×11-in. (216×279-mm) page images.

The image source is a helium-neon laser that is split into two beams to lay down two raster lines at a time on the belt, spaced to achieve 600-dpi resolution. A spinning polygon mirror focuses the beams on the belt. The beam erases the charge in each spot it hits, which prevents toner from adhering to the belt in that spot.

The toner is supplied from a cartridge, like all copy technologies, except that it uses a dual-component toner. The other component for the dual-component system is the carrier beads that attach to the toner until the charged belt attracts the toner.

One way the DocuTech achieves its fast speed is in the duplexed printing. Instead of flipping pages, the DocuTech uses software to determine the fastest way to accomplish the two-sided printing. It might for example, follow the printing of side one immediately with the printing of the other. In other cases, it could interleave the printing of the second side of one sheet with the printing of the first side of the next sheet.

Designed for long periods of uninterrupted service, the DocuTech is rated at one million impressions per month. During peak periods some report printing 2.5–3 million impressions per month.

A wide variety of paper stocks can be used. Paper weight can vary from 16-lb. (60-gsm) bond through 110-lb. (199-gsm) index. Transparency material, carbonless paper, both precut and full-cut tab stock, and recycled paper can be printed. Preprinted stock can be sent through the DocuTech if laser-approved, wax-free inks are used on the lithographic press. Before printing, the DocuTech checks to see whether the stock loaded in the trays matches the requirement specified on the job ticket. If the job ticket doesn't match, the system refuses to print the job. A unique solution to paper supplies is available from Roll Systems Corporation, whose DocuSheeter allows a DocuTech to operate off of a roll of paper. The 17-in.-wide (432-mm-wide) roll of paper is cut into sheets as it enters the printer, allowing for 12 hours of continuous operation at rated speed.

**In-Line Finishing Options**
All models of the DocuTech have stapling and two adjustable side stitches. This is performed in-line and has a maximum capacity of 70 sheets of 20-lb. (80-gsm) bond. The Model 135 and Model 6135 also have the capability of applying a heat-activated tape binding strip to the side of a book. The number of pages can be between 15 and 125, and can only be applied to 11-in.-long (279-mm-long) paper. The binding tape is supplied on a reel that contains 425 binds. The tape is available in black, blue, brown, gray, and white.

Following are some finishing options available for use in conjunction with the DocuTech. Chapter 15, which discusses bindery issues, also lists additional finishing options.

**Signature Booklet Maker.** The Signature Booklet Maker (SBM) is an optional extended attachment only for the DocuTech 135 and 6135. It is built by C.P. Bourg. The SBM serves an important role especially for products such as customized books. It takes finished signatures and produces saddle-stitched, folded, and trimmed booklets. In contrast, the in-line finisher, although it has two stitch heads, cannot saddle-stitch, fold, or trim.

The SBM can handle a variety of paper sizes up to 11×17-in. (279×432-mm) paper. The maximum thickness is equivalent to about 16 sheets of 20-lb. (80-gsm) bond paper. If a preprinted cover is needed, a Cover Insertion Module (CIM) can be added. This device allows a preprinted cover to be automatically inserted onto a pile of printed interiors before stitching.

**In-Line Perfect Binding.** The Bourg BB2005 in-line perfect binder from C.P. Bourg, Inc., is fully integrated mechanically and electronically to the printer's output. This binder works on the principle of hot-melt binding. Hot melts are 100%-solid thermoplastic adhesives. They are based on plastic polymers, compounded with suitable resins and other elements to give desired adhesion, viscosity, flow characteristics, and stability. Hot melts differ from other adhesives because they set by cooling rather than by absorption or evaporation of the liquid vehicle. This gives hot melts the instant bonding characteristics that is the chief reason for their popularity. Hot melt has a honey or pigmented white color.

The sheets are placed in the Bourg binder at the loading station and sent through the binding cycle. The binder jogs the pages, binds them, and ejects the book. The advantage of this binder is that it is available in-line with the Xerox DocuTech models. It is now possible to get a finished book with professional hot-melt binding in-line with the DocuTech printer.

The maximum book size that can be finished at this binder is 9×13 in. (229×330 mm), and the minimum book size is 7×10 in. (178×255 mm). The maximum book thickness is 1.6 in. (40 mm); roughly 350 sheets of 20-lb. (80-gsm) paper. The minimum book thickness depends upon the in-line printer speed.

**Off-Line Perfect Binder.** The Planax perfect binder, manufactured by Planax North America Inc., is an off-line high-quality PVA cold-emulsion adhesive binder used for binding books. (PVA — polyvinyl acetate — is a vinyl resin.) This equipment is used in making books that are printed by the DocuTech. In the Planax binder, the book block is locked into a traveling clamp. When a button is pushed, the book block is sent across a milling/slitting wheel, and then over the glue pot. The cover previously placed onto the cover mounting station is folded around the prepared and adhesive-coated spine.

The clamp is then released, the book block drops, and the clamp is sent back by pushing another button. The book is then removed and placed on a warm tray (about 140°F) to dry. On this equipment, it is possible to do slitting and milling up to ⅛ in. (3 mm) deep. The hourly production is about 100 per hour. Bound book blocks require a 15-min. dry time. The dry film coating of the adhesive should be between 0.0118 and 0.0196 in. (0.4–0.5 mm) in thickness.

The maximum bind size is 1.75 in. (44 mm) wide, 13.75 in. (349 mm) long, and 15 in. 381 mm) high.

**Book Station.** The Book Station is manufactured by Flesher Corporation, one of the business alliances of Xerox Corporation. The book blocks are perfect-bound at the Planax binder and finished to the final product at the Flesher station with covers, etc.

This equipment produces three-piece case books, which are hardcover-bound. This station is intended to produce short runs of high-quality hardcover books in a cost-effective and less time-consuming process. The short-run book assembly station guides the operator through a few simple steps. This is achieved by utilizing pre-assembled book blocks and hardcover components such as front and back covers, endpapers, etc.

This system features a precision spine support strip cutter and cover registration guides. The guides are easy to follow by the operator, and the whole process of completing a book can be accomplished in a matter of seconds. To be sure that the operator yields a square book, there is a cover alignment guide and a hinge creaser to yield professional cover joints. These features produce a hardcover-bound book that is creased at the hinge and opens flat.

**Off-line Adhesive Binder.** An off-line thermoplastic adhesive binder manufactured by California-based Powis-Parker is also available. It binds books from 3 pages to 350 pages, with a maximum thickness of 1.5 in. (38 mm). Documents that are bound are opened flat. Three binding strip sizes are available to suit thicknesses such as narrow, medium, and thick. All the binding strips are available in various colors to suit the customized preferences of the customer.

The fastback adhesive can be reheated to release and reseal pages. The cycle time (from placing a document in the binder to the finished binding) is approximately 25 seconds or less per book.

**Three-Edge Trimmer.** This three-edge single-knife trimmer is manufactured by Challenge Machinery Co., which specializes in the manufacture of graphic arts finishing equipment. The trimmer will cleanly finish the unbound edges of perfect-bound books. It follows a simple, safe process for finishing books. The paper-bound books are ready for packaging or for processing into hardback volumes. The pre-programming for standard stock sizes makes the trimming process simple and efficient. The book size is selected by toggling the size key on the keypad.

The DocuTrim resets itself and moves to load position. Following this step, the book to be trimmed can be inserted. The book is placed, bound side first, into the clamp with the top against the right guide. The foot pedal is pressed to clamp the book under low pressure.

The cut cycle automatically rotates the book and moves it to trim all three unbound edges. The cutting or trimming cycle can be interrupted at any time by pressing a foot pedal or cutting the infrared beams. After the third and final cut, the book is returned to the load position and the clamp is released. The trimmed book is then removed, and the process can be repeated for another book.

With this equipment, it is possible to do two-up trimming; for example 6×9 in. (152×229 mm) from 9×12 in. (229×304 mm). Its safety feature is its feed opening station, which is guarded by infrared electronic beams. The paper clamp functions under low pressure whenever the foot pedal is pressed and the beams are broken. When the operator's hand is removed and the infrared beams are intact, the clamp builds to maximum pressure and the cycle starts.

Another unique feature of this trimmer is the innovative design of holding the book at the clamp station. The book is held by the clamp at the spine without crushing or touching the spine edge. This way, all three edges of a book are trimmed without any crush marks.

The pile size is a maximum of 1.75 in. (44 mm), and book sizes of 6×9 in. (152×229 mm) to 8.5×11 in. (216×279 mm) can be trimmed with a maximum sheet size of 9×12.4 in. (229×315 mm). The book cycle time is approximately 30 seconds. The DocuTrim is available immediately for delivery off the shelf. All the service and support is provided by the manufacturer or distributor for this product.

# Chapter 24

# Kodak Lionheart

When desktop publishing technology first emerged in the 1980s, the tools and abilities were limited. There were no standard page description languages, file formats, and fonts, and it was before Windows arrived. Apple Macintosh computers owned the desktop publishing market, but with 512 KB of RAM and no hard disks there wasn't much you could do. At this point to do serious publishing, you needed heavy-duty software and hardware.

Like other manufacturers during this time, Kodak offered a closed or proprietary package that became the Kodak Ektaprint electronic printing system, or KEEPS for short. But as industry standards developed around IBM clones, the Apple Macintoshes, and Adobe PostScript, Kodak found KEEPS was locked out of the advances that others shared so freely. As customers leaned toward the open system, Kodak moved to industry standards and announced Lionheart in the fall of 1990. Today, Lionheart can work with documents created in any page description language, in any image file format, and from any kind of computer, and the software talks to, and links with, almost everything.

## Lionheart Architecture

The Lionheart system comes in many varieties. At its simplest, Lionheart architecture is a way to take computer-generated documents and output them to a high-speed printer. You can order a Lionheart product as a peripheral device, or as a complete system. One current offering, for example, is a package called the Lionheart ImageSource 92 printing system. It includes the Lionheart Print Server 75, print server software, a controller, and a 92-page-per-minute, 600-dpi printer with finisher. That's everything you need to do some pretty sophisticated black-and-white on-demand printing.

The Lionheart architecture is made up of three distinct components:
- A network strategy for document management
- A family of printers
- A family of enabling software

In a network, the Lionheart server and software can function as a central resource for document creation and distribution. A user can create a document with input from employees in several departments, using both IBM and Macintosh platforms. Typi-

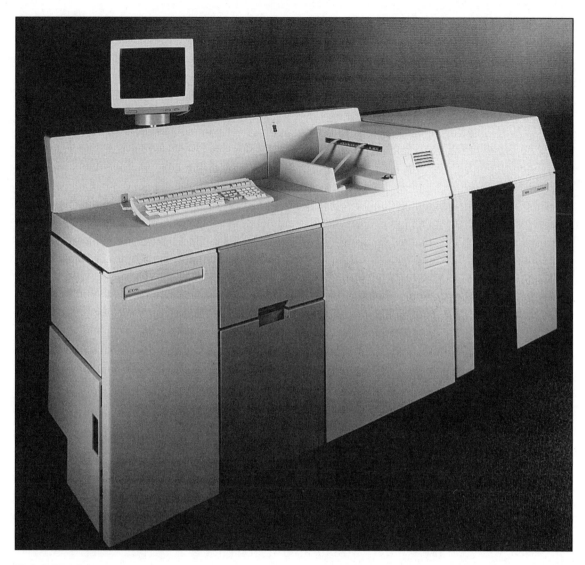

Kodak Lionheart.
*Courtesy Eastman Kodak Co.*

cally the user writes some of the report, and he or she picks up documents from other sources.

The user can add drawings by scanning them into the system, and the software assembles those pieces into a single electronic document, routing them to any Post-Script printer on the network. Most Lionheart users configure their system specifically for one of Kodak's 92-page-per-minute printers.

With Lionheart, users can post a document on the network for others to pick up, or send a technical manual to a CD writer for output on CD-ROM discs, because the system reaches beyond the paper-bound world. Sometimes the best way to distribute a document isn't in hard copy at all. The system supports output to microfilm or microfiche, and documents can be offered on more than one medium (both as hard copy and as a CD-ROM disc, for example).

## Printers

The ImageSource 92 is the latest Lionheart printer. The first printer that Kodak introduced (in 1990) was the Kodak 1392 printer, with a top speed of 92 pages per minute and a resolution of 300 dpi. The 1392 printer (which has been upgraded over the years to its current version, the model 44) remains one of the more popular Lionheart printers sold. It's a reliable workhorse — rated for up to 1.5 million impressions per month.

Other Kodak printers sold as Lionheart components include the:
- **Kodak ImageSource 70 copier-printer,** a 400-dpi, mid-volume specialty printer. Though slower than the workhorse 1392 and ImageSource 92 printers, it provides "Free-Form Color printing" (discussed later), where up to three colors can be imaged, in addition to black. Users can also feed a hard copy through its document feeder once to scan a document in, then print multiple copies from memory. It also features Kodak's high-definition imaging (HDI) technology (discussed later).
- **Kodak ColorEdge 1560 and 1565 copier-printers,** which offer full-color, 400-dpi laser printing at a speed of up to 7 ppm and are rated for up to 15,000 impressions per month. They accept paper stock up to 11×17 in. Paper weight can be 20–28 lbs. in the drawer, or up to 60-lb. cover (90-lb. index) in the stack bypass. The 1565 offers auto duplexing, and an optional film scanner allows copying/scanning of slides. In order for them to function as network printers, you also need a color server.

The Lionheart software and server will also support a variety of other networked PostScript printers. Kodak will customize the server software to accommodate any number of network devices.

## Software

In addition to the basic print server software outlined above, Lionheart software is available with other functions. Each module gives the user different capabilities.

The first category of software module is value-added printing applications. File Merge software allows users to build compound documents as described earlier, combining files from various sources (including other network users). Info Merge software combines variable information with standard text and/or layouts.

The scan-store-print software lets users convert hard copy documents into electronic images for on-demand printing. Image Merge software combines scanned documents (stored as TIFF Group IV) into an electronic document. Image Library Management software maintains an indexed collection of scanned images and lets users access them freely. Image Edit software allows users to cut, paste, crop, rotate, and reverse images prior to printing or merging with other documents.

How do all of these pieces come together? A user can assemble a report from word processing files, scanned images, page layout files, spreadsheet files, ad infinitum. Along the way, the user accesses a few pages of boilerplate on a new product and scanned images of that product, which were stored on the print server's hard disk. They are cropped and scaled using image editing software. When the whole package is complete, the user posts it to the server for printing, along with instructions—what network printer to use, what kind of paper, binding, delivery, and even which company account to bill.

Kodak promotes its ability to offer customized software. One government agency that distributes large volumes of its own publications of regulations, for example, uses Lionheart and a customized on-demand printing software module. Customer service representatives (CSRs) take phone calls from the public, who request specific publications. The CSRs then note which publications the caller wants, and enter the caller's name and address. The job is then released to the Lionheart system, which prints all requested publications, along with a cover sheet that acts as a mailing label for shipping.

## Modular Migration

Users who move beyond the basic server/software/printer configuration to build a networked printing solution for their company may find that not every part of the system bears the Kodak name. A company that already has a Sun SPARCstation 20, for example, may choose to buy the software without the server. It will usually get one of Kodak's high-volume printers, too; otherwise, it will miss out on much of the software's value.

As the company grows, it may choose to build a system by starting with one printer, adding another (perhaps with Free-Form Color printing or full-color capabilities). Beginning with version 4.0 of the Lionheart print server software, the system is designed to work with existing generic PostScript printers. Therefore, a company may decide to link an existing, third-party PostScript printer to the Lionheart system, or even a whole departmental local-area network.

Lionheart has, by and large, avoided the necessity of "forklifting" out the old unit and installing a new one. From the first 1392 printer to the current 1392, model 44, users have been able to upgrade their printer by swapping old internal components for new ones. The whole cabinet doesn't have to come out of the building.

## ImageSource 92 Printing System

Lionheart is available in several configurations. One is the ImageSource 92 printing system. It combines print server hardware and software, a controller, and a high-volume, black-and-white laser printer. The Kodak ImageSource 92 printer is a 600-dpi printer that offers the appearance of 120-line screens and a top speed of 92 ppm. It is rated for a capacity of up to 1.5 million impressions per month.

The printer's internal print controller is a 66-MHz Intel 486 CPU with 4 MB of RAM, GGX and TIFF accelerator cards, an Ethernet card, and a 2.1-GB hard disk. The disk is capable of processing up to 12,000 TIFF Group IV pages, at an average of 150 KB per page. It accepts legal, letter, and A4 paper, in weights from 20- to 110-lb. bond. It prints in both single-sided and duplex modes in a single pass.

In duplex printing, side one is imaged and the paper is flipped by a vacuum turnover drum. Then side two is placed on the paper. The completed page is transported over a cushion of air, and both sides of the paper are fused simultaneously. The entire process takes place in a 27-in. paper path, at full machine-rated speed.

An on-board print controller manages page printing of images produced by the interpreter software. This software processes job header files and TIFF Group IV images, and temporarily stores them on the controller's internal hard disk during the imaging process. Communications between the controller and the print server are accomplished by a dedicated Ethernet link using PC/NFS.

The Print Server 75 is a Sun SPARCstation 20, operating at a 75-MHz clock speed with 32-bit multiple processors, internal CD drive, 3.05-GB hard drive, and 8-mm tape drive. Standard interfaces include 10-megabit Ethernet, 10-megabit SCSI and SCSI-2, two RS-232C/RS423 serial ports, a Centronics parallel port, and four expansion slots.

The server comes equipped with a software RIP (raster image processor) that utilizes the PostScript Level 2 implementation of Adobe's Configurable PostScript Interpreter (CPSI). The interpreter software can process up to 999 PostScript jobs while the printer is running. The interpreter software converts each page to a TIFF Group IV image.

One of the main reasons for utilizing a software-based RIP is to keep up with the most current hardware capabilities. The faster the workstation, the faster the conversion from PostScript to TIFF Group IV. The Lionheart Print Server 75 is, on average, 25% faster than its predecessor, Kodak's DL Power Server. Lionheart architecture also provides the capability to simply pass a print job through the server without Post-Script interpretation if you want to let the printer handle the ripping.

Lionheart Print Server software 4.0, which resides on the server, allows a user to view and modify the electronic "print ticket" that accompanies each job. The user can assign jobs to different print queues (sending them to different printers, or changing the order in which they'll be printed) and archive jobs for reprinting in the future.

## Color Servers for Color Printers

Kodak ColorEdge 1560 and 1565 copier-printers require a color server to accept and process network print requests. Seven models of the Fiery XJK color server are available. The recently released Fiery Color Server XJK 1300, XJK 2600, XJK 3900, and XJK 5200 offer faster processing and more accurate color calibration than previous models. These models offer 128–512 MB of RAM. Large RAM memory and a MIPS R4600 or R4700/133-MHz processor speed page buffering of large files.

Each of these servers is packaged with a Fiery Color Server XJK Command Workstation. The Pentium-based workstation adds convenience and raises productivity by allowing customers to change files and perform calibration at the workstation. The workstation uses a Windows 95 Operating System and comes loaded with EFI Fiery utilities. To facilitate network management, the command workstation contains an 850-MB hard drive, 65 fonts, and 16 MB of RAM.

The command workstation's print calibrator and Kodak color management software provide consistent color rendition. Users can control an additional printer from the same server.

Fiery XJK, XJK 100+ and XJK 200+ color servers offer 48–208 MB of RAM. However, through some magic performed by the XJ chip set, effective page buffer RAM is increased. All three models come with a 100-MHz R4600 processor, with a 16-KB data cache. A 540-MB internal hard disk is standard, but it can be upgraded to 1 or 2 GB.

## Full-Color and Free-Form Printing

A wide range of printing falls between simple black-on-white and full color. Documents have been designed for years with headlines, some text, and boxed areas in a second or third "accent" color, and Kodak has served this market with its 1580 copier-printer and several copiers. This method was relatively imprecise, however, since an operator had to designate the area to be colorized using a stylus and digitizing tablet. It has also been difficult to achieve consistent color registration.

More recently, document creators have been able to designate within the electronic document file up to three colors plus black. Kodak calls this Free-Form Color printing. With Free-Form Color printing, users can import computer-generated color graphics, and the printer will output the file using the on-board color that it determines is closest to the one in the file. Free-Form Color printing is available on the new ImageSource 70 copier-printer, which offers black plus three additional colors on-line. The color palette includes red, yellow, green, and blue.

Even with four colors of toner, these printers can't offer traditional "process" color (the process through which printers have long reproduced color photos by overlapping four colors of ink or toner dots). What they offer instead is a way to produce business graphics at a much lower cost than with a full-color electronic printer. Print speeds vary as the number of colors on a page increases: single-color (black) printing is rated at up to 70 pages per minute; two-color printing is rated at up to 35 ppm; three-color at up to 23 ppm; and four-color at up to 17 ppm.

## LED Imaging Technology

Kodak utilizes LED (light-emitting diode) print heads in the 1392, 3072, ImageSource 70, and ImageSource 92 printers. An array of LEDs marks a photoconductive film belt by emitting red light. The ImageSource 92 printer's LED print head has 8448 LEDs in a single, 14-in.-long line. The print controller transmits bitmapped information to the print head and triggers the array. Individual LEDs are then lit or left dark, depending on the bitmap information for that line of the image.

After the print head images a line, the film belt advances. The LED print head remains stationary. That means there's no wear, no vibration, and no wobble in the print head.

It also lasts longer than the print head in laser printers. Why? High-volume lasers mark the film belt for all white areas on a page (the areas that won't receive toner). In the LED printer, though, the process is reversed. Only the black areas are marked. Most printed pages have much more white than black, so marking the black areas requires fewer dots. That means the LEDs are not on very often, which extends their life.

Kodak also claims that LED printers are sharper than laser printers with the same nominal resolution (dpi). The reason is the toner. In order to mark only the black, Kodak developed a special toner that is attracted to the marked areas of the film, rather than the unmarked ones. The toner has a negative charge and produces small, uniform dots.

Under a magnifier, a laser-printed character on a piece of paper has tiny marks of stray toner particles around each character. These tiny marks are called "satellite particles." The more satellite particles, the more ragged the image appears. The satellite area around the LED dot is much smaller than that around a laser dot — due to the negatively charged toner.

The new ImageSource printers sport a further refinement in image quality, which Kodak calls "high-definition imaging." HDI refinements include using finer toner par-

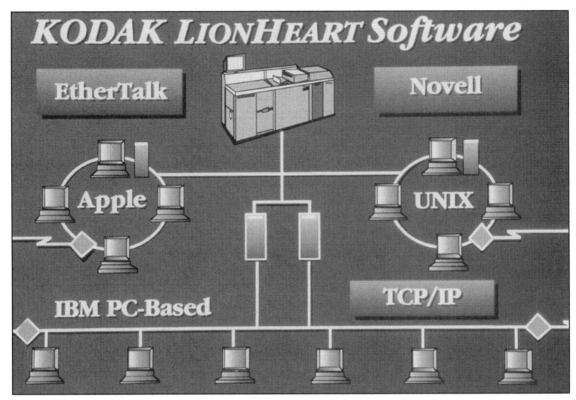

Lionheart networking options.
*Courtesy of Eastman Kodak Co.*

ticles, a new developer, and an improved film loop. Together, they improve reproduction of halftones and fine line artwork.

## Finishing Options

ImageSource 70 copier-printers can fold and staple output to produce finished 8½×11-in. or 5½×8½-in. booklets, from stock up to 11×17 in. in size, including special colored or heavier-weight cover stock. The high-volume printers (and ColorEdge copier-printers) offer limited finishing capabilities. Covers can be printed on a separate type of paper, and completed documents can be stapled in various positions. To add to the range of finishing options, Kodak offers a third-party finisher from Duplo that can be configured for on-line finishing to produce booklets from the 1392 or ImageSource 92 printers.

## Advantages

While other printers have added PostScript interpreters, one of the major advantages of the Kodak system is its adherence to industry standards. Two industry standards are PostScript and open network architecture. The Lionheart server on a local-area network, with the appropriate software, lets users manage on-demand printing just as a print shop manager would manage traditional lithography. Documents can be routed to the printer that is most appropriate for the job. Documents with the same characteristics — such as black printing on white stock, with blue covers — can be routed to the same print queue.

The Lionheart system is compatible with Macintosh and PC-compatible desktop computers, Sun workstations, and DEC VAX computers running under VMS. It supports EtherTalk, TCP/IP, and Novell networks. Lionheart print drivers are available for most desktop publishing programs, image creation/manipulation programs, and the most popular word processors.

The system also supports a variety of page description languages and image file formats. This includes Adobe PostScript, levels 1 and 2; Hewlett-Packard's PCL 4 and 5 page description languages; and TIFF and CCITT Group IV image files.

## Disadvantages

Not every printer is equally flexible, however. The 1392 printers support all the standards. The ImageSource 92 printer, on the other hand, supports PostScript 1 and 2, and TIFF. The ImageSource 70 printers support PostScript 1 and 2.

Another issue is matching your primary applications to the system. Each of the printing systems outlined in this book was designed first to fill a particular niche. If you primarily need to print full-color documents, the Lionheart system (and Kodak ColorEdge copier-printers) won't be fast enough to compete with systems designed for full-color work, like Indigo, Chromapress, and Xeikon. For black-and-white, 8½×11-in. work, however, Lionheart software and the 1392 or ImageSource 92 printer are a great match. They offer lower cost and higher productivity.

Kodak also lacks a high-volume 11×17-in. printing solution. The ImageSource 70 printers handle the format, but at a slower speed (70 ppm), and they are rated for lower volumes (200,000 impressions per month). So if 90% of your work requires 11×17-in. printing, look elsewhere. If 10% of your work requires 11×17 in., you might find that the mid-volume printers will serve you well.

## Future

Over 30% of current Lionheart system users employ the system to produce black-and-white manuals from PostScript files. Most of that base was established with the 300-dpi Kodak 1392 printer. With the higher-resolution ImageSource 92 printer, and various full-color and Free-Form Color printing options now available, the Lionheart system is becoming more attractive for other demand printing applications, including those with more halftones and color.

The Lionheart architecture also has the power to handle what Kodak calls "distributed" printing, for organizations large enough to take advantage of it. Until now, on-demand printing has been viewed as a function for the central reproduction department. Lionheart software breaks the boundaries. A company may start today with Lionheart driving an ImageSource 92 on a local-area network at its Chicago headquarters. But next year, it may have other high-volume printers on the network — by now a wide-area network — including some in Los Angeles and New York. The company can issue updated documents to all three cities over the network, and print them simultaneously.

# Chapter 25

# Color Copier/Printers

Color copiers are essentially a scanner at the top and a digital printer at the bottom. They are designed for making one or several copies of spot- or four-color process subjects. When controlled by computers they can be used for very short-run color printing (<100–500). The first color copier used for digital color printing was the Canon CLC. It was also the first copier device to integrate a PostScript controller. The Canon Color Laser Copier is an analog copier with a scanner that color-separates the color original into the four separation colors (CMYK), each with 400-dpi resolution and 256 levels of color.

It produces the four composite toned images on the paper mounted on a drum, after which the images are fused on the paper using a special fusing oil and heat. A special image processing unit (IPU) is used to produce images digitally from PostScript files.

## Review

Today's color copiers use the same digital technology as scanners and printers, but face greater issues. Color copiers are inherently more expensive because the images are "built up" during each pass with the toner color. Depending on the specific manufacturer either three or four colorants can be used, or a material such as a photographic material can be imaged in a single pass.

Most of the less-expensive or office-based copiers that use the toner build-up scheme use cyan, magenta, and yellow toners. Theoretically, only three colors are required to print. However in traditional printing and color copying, adding a fourth color, usually black, improves picture quality and decreases cost for printed text.

Color copiers often use successive scans to estimate the amount of toner of each color to apply. This adds complexity to the printing because each pass must maintain tight register. Another issue with color copying is the ability to mimic shading. This is not as much of an issue with black-and-white copiers because people don't expect the picture to copy that well. Generally, the darker the text the more the customer is satisfied. With color copiers, lightness is a critical issue in the reproduction of color images. Correct lightness depends not only on the copier's scanning or optical systems to capture the correct color, but also the copier's print engine.

## Optical or Digital?

Digital copiers have advantages over copiers with optical systems. Digital copiers use a scanner that reads the color and then creates the image with either a laser (as in

Canon's CLC 300 and CLC 500) or a thermal head (as in Toshiba's ChromaTouch 1000). Another advantage of the digital copiers is the ability to be connected to a computer to output files directly.

Optical copiers, in contrast, flash light and record light as it is reflected from the original through three or four color filters onto a photosensitive drum. Although not as flexible as digital copiers, optical copiers are generally less expensive and have the advantage of copying three-dimensional objects, a feature important to the jewelry industry.

Ricoh's NC100 is one of the more advanced optical color copiers because of its image editing features. It has capabilities such as color and area identification features that allow users to change a single color on the original to another color, or to remove the color on the copies.

### Color Copier Interfaces

Regardless if you own a color copier today or you're considering the purchase, a key issue today is connectivity. Connectivity, or the ability to connect your computer to the copier, is more formally called a color copier interface, or printer controller.

These interfaces can include a host of other capabilities such as scanning from the copier into the computer, print spooling and job queue management, image editing, color proofing, and printing. According to market research studies by BIS Strategic Decisions, 60% or more of all color copiers sold in the U.S. in 1993 included digital interfaces.

Companies offering products in this category include Agfa Graphics Systems, ColorAge Inc., COLORBUS, Inc., DICE America, Electronics for Imaging, Inc. (EFI), and User Friendly Operating Systems. The market leader today is EFI, claiming sales of over 10,000 of the Fiery controller.

Some interfaces also offer print spooling functionality. Once installed on an AppleTalk, Ethernet, or NetWare network, these products allow users to access the copier through either the Windows Print dialog box or as a Macintosh chooser option.

### On-Demand Printing from Copiers

Combining all the features, you can produce high-quality (400×400-dpi), full-color (CMYK), large-format (11×17-in.) printouts on-demand and in quantity (typical runs range from 100 to 500 copies). The issues are speed and cost.

The first Canon color copier users were able to price their 8.5×11-in. color copies at $3.00, and the 11×17-in. copies at $6.00. Today, the standard page is going for under a dollar (mostly about $0.75), and the double-page sheet is going for $1.50 to $2.00. The introduction of the EFI Fiery with its superior interface to PostScript essentially halved the price. Thus, customers could go directly from electronic files to color pages.

Today most of the copiers are plain-paper, high-volume machines (from 5 to 7.5 pages per minute), and the cost per page can be as low as $0.10–0.25, or $0.30–0.50 including leasing and service contracts.

As a result, there is a pent-up demand for digital color printing. It is based on three major factors:

- **The number of pages in electronic form.** Counting both PostScript and pages produced on color prepress systems, over half of all pages are now in digital form and thus ripe for electronic printout.

- **The preponderance of color printers and copiers.** As a result of their popularity, many buyers have been exposed to color reproduction. They have also developed an acceptance for color printer and copier quality levels.
- **The cost pressures on American business.** Without an expanding economy, business is maintaining profit levels by cutting cost. Just-in-time approaches help. But for JIT to work, it must also offer a time advantage as well as a cost advantage.

We estimate that almost 1.14 trillion pages worldwide are available for digital printing. This volume will come from:
- Volume that would have gone to color printers and color copiers   19%
- Volume that would have gone to conventional color printing   45%
- Volume that would have been black-and-white copying/printing   15%
- New volume developed for digital color printing   21%

Because the bulk of volume will initially come from commercial printing, commercial printers have become the first users.

At $1.00 per standard page, it becomes a $1 trillion market. We estimate that the cost to produce that page with equipment, personnel, supplies and service would be in the $0.40–0.60 range, providing a reasonable return for the equipment user. Since there will be no other source for the toner and ink, users will consider these costs carefully.

## Canon's CLC 700 and CLC 800 Color Copiers

During a press conference at the Seybold San Francisco conference in 1995, Canon USA unveiled the next generation of its Color Laser Copier series along with a new PostScript Level 2 controller based on a Silicon Graphics workstation. Two other vendors, COLORBUS and ColorAge, have also announced new versions of their controllers that can drive multiple copiers.

Canon's copiers, the CLC 700 and CLC 800, feature a new laser engine with duplexing (two-sided printing) capabilities and output speed of 7 pages per minute, 40% faster than the current generation. The copiers also have numerous features designed to enhance image quality, including a smaller laser beam spot, a new method for sending charges to the transfer drum, and an anti-moiré filter. The CLC 700 also features manual duplexing capability, while the CLC 800 offers automatic duplexing. Both copiers use a new, larger drum that yields 40,000 copies as opposed to 20,000 in the old drum. They also feature a 50-sheet stack bypass that allows use of transparencies and paper stocks up to 90-lb. (163 gsm) index or 60-lb. (162 gsm) cover.

Presenting the copiers, Canon's Jim Sharp said that the company performed extensive market research to discover what features and capabilities current CLC customers — and users of competing products — were looking for. These included better image quality, duplex printing capabilities, faster output speeds, and the ability to print on a wider range of paper stocks. He stated that Canon currently has almost 70% of the market for digital color copiers and hopes to increase market share with the release of the new models.

Suggested retail prices are $33,000 for the CLC 700, which has a duty cycle of 10,000 copies per month, and $49,900 for the CLC 800, which can produce 20,000

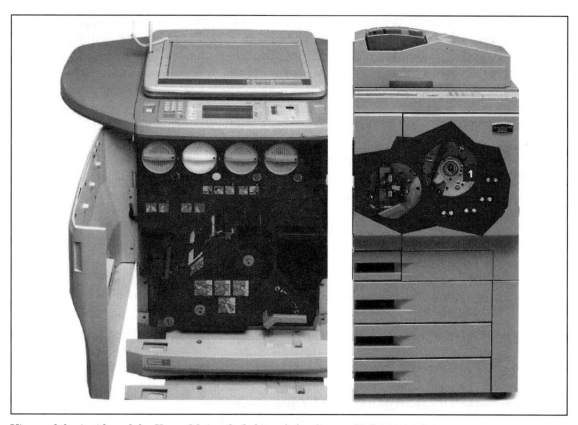

Views of the insides of the Xerox Majestik (left) and the Canon CLC 800 (right).
*Courtesy of Xerox Corporation (left) and Canon U.S.A. (right)*

copies per month. An optional Film Projector/Film Scanner III, priced at $15,000, allows the copiers to produce output from slides or transparencies up to 4×5 in.

Perhaps more significant than the new copiers was the announcement of the ColorPASS-1000 controller, which was developed jointly by Canon, Adobe Systems, and Silicon Graphics Inc. The controller, which will compete with products from EFI, ColorAge, COLORBUS, and other vendors, runs on an Indy workstation from Silicon Graphics and features an Adobe PostScript Level 2 interpreter. In addition to supporting the new CLC 700 and CLC 800, the controller is also compatible with all earlier CLC models and will support future models as well, Canon says.

## Canon CLC 1000 vs. the Xerox DocuColor and Scitex Spontane

The CLC 1000 was secretly shown a few years ago and then carted out for trade shows and the like. Canon is now taking orders and says shipments will start during the first quarter of 1997. Xerox DocuColors and Scitex Spontanes are being shipped at present. Ideal Printing & Engraving in New York City has the first DocuColor 40; however, Sonic Printing has had a Spontane for a few months. We expect about 500 of the DocuColor/Spontanes worldwide by the end of 1996.

It is interesting that Xerox chose a graphic arts firm for its first site, given that Scitex was to sell into that market. We guess that it makes no difference and may the best marketer win. The Spontane and DocuColor list for $200,000. The CLC 1000

Kodak 1560 copier-printer (left) and Kodak 1565 copier-printer (right).
*Courtesy of Eastman Kodak Co.*

pricing had been a moveable feast but appears to be in the area of $75,000 without a RIP. The Xerox DocuColor and Scitex Spontane (which we abbreviate as XDSS) is about $130,000 without the RIP.

The major difference between the Canon and the XDSS is speed: 31 ppm vs. 40 ppm. But when it comes to duplexing, they are almost equal. Canon uses a duplex tray to hold copies and then feeds them back through; the XDSS flips each sheet and duplexes one at a time. The result is 14–15 ppm for both.

The CLC comes with a standard 2,500 sheet high-capacity paper unit. The XDSS can hold 1,500 sheets if you use every paper drawer.

## Color Copier Numbers

At a press conference, Canon showed a very quick chart of its market share. The chart indicated that Canon had 68.2% of the color copier market in the U.S. and that Xerox had 11.7%. That means that all other color copiers must add up to 19.1%. They showed U.S. installations in 1992 at 6,900 and 1994 at 8,700 (probably a projection). So we tried to extrapolate from these clues what the numbers might look like:

| Year | U.S. Units | | |
|------|-------|-------|-------|
| | **Canon** | **Xerox** | **Other** |
| 1994 | 8,700 | 3,000 | 5,200 |
| 1993 | 7,300 | 2,000 | 4,500 |
| 1992 | 6,900 | 1,000 | 500 |
| 1991 | 5,800 | | |
| 1990 | 4,000 | | |
| 1989 | 2,000 | | |
| **Total** | **34,700** | **6,000** | **10,200** |

BIS Strategic Directions has estimated the Canon population at 29,000 to the end of 1993. We assume that the units sold to Kodak are counted as Canon's sales. But are they also included in "Other" as a sale by Kodak? We do not know. What we may know is that the U.S. population of color copiers is around 50,000 units, and each is potentially a low-level digital printer.

### Scitex Spontane and Xerox DocuColor Establish New Digital Demand Printing Market

Introduced in Japan in 1995 as the 4040, introduced in May 1995 at Drupa as the Scitex Spontane, and introduced in May 1996 as the Xerox DocuColor, this digital printing system incorporates a high-speed full-color print engine designed for high-quality color reproduction of flyers, brochures, pamphlets, comprehensive proofs, and short-run, print-on-demand applications. A feature for variable information that allows printing localized inserts, mailers and flyers, and customized catalogs, tailored to interests of targeted readers is optional with most RIPs.

With a small footprint and a speed of up to 40 full-color A4 pages per minute, the Xerox DocuColor and Scitex Spontane (XDSS) provide an entry-level solution for printers, repro houses, and digital service bureaus interested in making the evolutionary shift to digital printing.

The XDSS printing system integrates seamlessly into digital environments, allowing automation and management of the entire workflow, saving labor and consumable costs associated with other printing processes. The XDSS accepts files in PostScript, Scitex, and other digital formats, processes them, and prints them in CMYK, in A6 to A3 formats, including automatic duplex. Special features include electronic collation of documents, and three different paper trays that can be loaded with variable stock and material and integrated automatically into the final documents. The printing priority queue can be altered for rush jobs or quick proofing.

The XDSS results from a strategic alliance with Xerox Corporation, Scitex, and RIP suppliers to jointly develop a series of highly productive solutions for integrated color documents, bringing together their accumulated expertise in the relevant technologies and markets. The companies will join efforts in the development and distribution of products for the on-demand digital printing market. Shipments began near the end of 1995.

Scitex demonstrated a Fuji-Xerox printer-copier connected to a Scitex RIP-server at Drupa in May 1995. The engine is manufactured by Fuji-Xerox in Japan and sold by both Scitex and Xerox. The device is a developmental extension of the Majestik print engine with four imaging and toning stations in a row. It was first prototyped back in 1993–1994, but Xerox decided not to go forward with it as a product. The folks at Fuji persevered and that led first to the Scitex relationship and the Spontane and then to a Xerox version. Resolution is 400 dpi with 8 bits per pixel, which makes for good looking prints.

**A Simple Design.** Japanese market watchers have mentioned a machine named 4040 — 40 simplex 4-color pages per minute is the XDSS speed — which is the name most Europeans also know it as. This speed puts it at 900 duplexed pages per hour, just under the 1,000 for the Indigo and way under the 2,100 for the Agfa-AM-IBM-Xeikon. The quality is just about at the Agfa version of the Xeikon, which is the best of all of them, even though they use the same engine. The quality is below Indigo, which is almost equal to offset lithography.

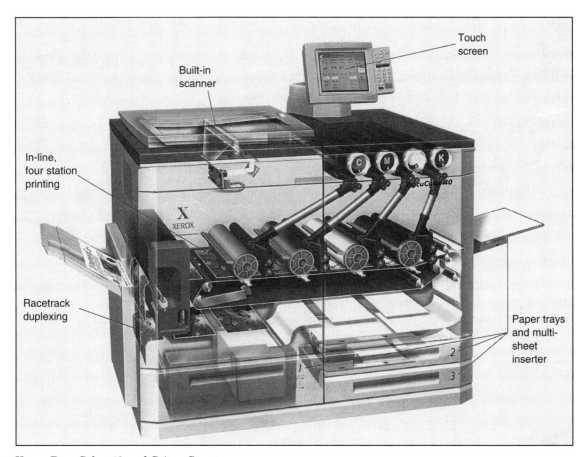

Built-in scanner

Touch screen

In-line, four station printing

X⁴ XEROX

C M K

Racetrack duplexing

Paper trays and multi-sheet inserter

Xerox DocuColor 40 and Scitex Spontane.

Paper is moved via a clear plastic belt that holds it with static electricity as the paper passes beneath four laser and toning heads. A unique "flipper" mechanism turns the paper over so it can then make its way back through the printing path to print the flip side. Duplex printing cuts the speed to 15 pages per minute — that was what we timed on the XDSS. To maintain this speed, the system prints the first side of a sheet as the previous sheet is being flipped and re-printed.

**Printer or Copier?** Xerox and Scitex have introduced their versions as a copier and as a printer. The copier configuration is priced at $130,000. Cost per copy on the XDSS, using our cost model, is about $0.38 per 8.5×11-in. (216×279-mm) sheet, printed four colors over four colors, not including paper and RIP time. This was based on some assumptions about toner pricing, which are now solidifying as XDSS users have actual field experience. Consumables have the same pricing, and the toner bottles and receptacles are the same. Thus the cost per print unit should be the same for both companies.

**Options.** A collator-stacker would be a worthwhile option as would an automatic document feeder. These have not yet been announced. There could be others in terms of on-board memory and finishing options. The major option of interest to digital color users is the RIP. The XDSS also cries for a high-capacity sheet feeder, which is promised shortly. At 40 ppm, you go through a lot of paper.

EFI Fiery and other RIPs are coming. Prices on the RIPs would probably follow existing trends in the range of $40,000–80,000, which would bring the total system near $200,000, about half of an Indigo or Xeikon.

## Markets

There is a difference in their markets. Scitex is very strong in prepress and printing services, although one-third of DocuTechs are now in printing companies. Xerox is strong in corporate markets, although some of its first big orders may be from printing firms. It will be interesting to see what happens when the two compete for the same account. It would be like Dole running a negative ad against Dole. Like DocuTech, Spontane scans or inputs pages into memory for a major advantage in speed. We do not think that this device is the illusory color DocuTech. It is, however, a serious color product.

Our guess is that the first users are shops with multiple Canon CLC color copiers. Most are rated at 5–7 pages per minute. It is easy to see how a machine like the XDSS could easily replace three to four color copiers and result in a reduced maintenance contract rate and faster turnaround potential.

| CLC 1000 vs. DocuColor/Spontane | | |
|---|---|---|
| | **CLC 1000** | **DocuColor/Spontane** |
| 8.5×11-in. simplex pages per hour | 31 | 40 |
| 8.5×11-in. duplex pages per hour | 14 | 15 |
| Paper capacity (without bypass) | 5,200 | 1,500 |
| Price without RIP | $75,000 | $130,000 |
| RIP | $40,000–70,000 | $62,000–75,000 |
| Street price with RIP | $120,000 | $169,000 |

Scitex and Xerox now have the advantage because they are shipping. Canon could reclaim lost ground because of price. The battle in the new middle market of color digital printing has only just begun.

# Chapter 26

# Hybrid/Presstek Technologies

According to the the *American Heritage® Dictionary,* Third Edition the word "hybrid" means "Something of mixed origin or composition or something, such as a computer or power plant, having two kinds of components that produce the same or similar

Heidelberg GTO-DI sheetfed press that uses the Presstek direct-to-plate imaging technology. *Courtesy Presstek, Inc.*

results." We use the term "hybrid" when discussing the Presstek-based presses because they combine traditional press technology with direct-to-plate technology.

Introduced at PRINT 91, the first hybrid printing press was the Heidelberg GTO-DI. It was jointly developed by Heidelberg and Presstek (Hudson, NH) and used spark discharge technology to burn the plates on the press. According to our definitions, it clearly is a digital press because you can print out digital files. Whether it is an on-demand press is controversial. Some people feel that it is more accurate to refer to the GTO-DI as a computer-to-plate or computer-to-press device. According to our definition, if you print files quickly and efficiently in short runs then it qualifies as an on-demand press. Others argue that an on-demand press should be able to customize each page. Regardless of how it is characterized, it is important to discuss.

## Heidelberg GTO-DI

The first ablation process for platemaking was Lasergraph introduced in 1974 for the production of letterpress relief newspaper plates from digital information using high-power $CO_2$ lasers to etch the images on copper-plated plastic-coated metal-based plates.

The first successful hybrid press or re-engineered conventional offset lithographic press that allowed the plate to be imaged directly on the press was made by Presstek and Heidelberg in 1991. Presstek developed a direct-to-plate offset printing technology for short-run, low-cost color reproduction, which Heidelberg calls "Direct Imaging." The letters "DI" are used to designate presses with this capability.

The first device was actually built from scratch with a common impression cylinder and four sets of plates and blanket cylinders around the circumference. But the relationship with Heidelberg moved toward adapting an existing press rather than designing one from the ground up.

Thus, the first DI device used a popular four-color press, the 14×20-in. (356×508-mm) Heidelberg GTO, which was modified to image plates after they were mounted on the press. As a result, the four separations are in close registration (Presstek claimed a corner-to-corner error tolerance of ±0.0005 in., or ±0.013 mm). This technology resulted in shorter makeready times.

The Heidelberg GTO-DI merges a traditional printing press design with the latest in on-line, digital plate production solutions. The key to the success of the GTO-DI and Presstek's Direct Imaging lies in the reduction in prepress and plating processes, which can account for up to 80% of the time and cost to print in run lengths under 5,000 impressions.

This turned out to be a major technological breakthrough for Direct Imaging because it marked the first time that a plate could be imaged in daylight without chemicals. This revolutionary printing plate had a silicone surface and thus required no water to control the ink. It was originally imaged by a spark discharge technology that essentially drilled a hole into the plate surface, acting almost like gravure printing.

The plate exposure system operated at a fixed resolution of 1,016 dpi, which is sufficient for text and for images at screen rulings up to 100 lpi, with standard halftone generation methods. The plates used by Presstek comprised a three-layer sandwich: a half-mil-thick silicone surface, an aluminum ground plane, and a Mylar (polyester) base. The exposure of the plate was based on the spark technology, which is a technology that uses tungsten-needle electrodes and electric current to image a dot on the plate.

The other important issue in Presstek's first version was the spark system. An array of 16 needle-sharp tungsten wires, positioned at a height of two mils above the plate, delivers the spark to the plate. The array is not at exact angles to the plate's rotation, so each needle makes its own track and is fed from its own data stream. The tracks are less than a thousand of an inch apart (1/1,016 in. specifically), although this could easily be changed to support higher resolutions.

A spark is like a miniature lightning bolt. To minimize the dot-to-dot variations, Presstek does everything it can to keep the spark gap in a constant environment. Since it is vital to keep the spark gap from clogging up with silicone chaff, Presstek brushed away the chaff with a small rotating brush.

Secondly, the air was filtered and bubbled through water to humidify it. The air was directed at the plate through nozzles between the electrodes. This minimized and stabilized the break-even voltage of the spark gap. Thirdly, since it is important to keep the electrode array at a 2-mil spacing (cylinders are not perfectly round), Presstek measured the altitude and adjusted it on the fly.

The clever measuring apparatus took advantage of the air nozzle that blows filtered air toward the plate. A pressure sensor measures the back pressure of the air flow. If the distance from the plate were to increase for any reason, the back pressure would fall. The control electronics would react by changing the current in a small galvanometer motor attached to the cam that sets the recording head's height. However, the tungsten needles are gradually consumed by the sparks they emit. Presstek packed a replacement array for every ten plates.

## Pearl Technology

In the summer of 1993, Presstek announced that it had achieved a substantial advance in the plate recording technology by switching from the spark-erosion approach to a laser diode approach. Presstek called its new platemaking system "Pearl." Unlike the spark technology, which was funded by Heidelberg in exchange for exclusive marketing rights, the new Pearl technology was funded entirely by Presstek. As a result, Heidelberg's license for Pearl was non-exclusive.

The introduction of the Pearl technology overcame the quality issue. The Pearl upgrade replaced the spark discharge technology with laser beam technology. The laser technology works analogous to printing film from a PostScript imagesetter.

The new imaging system still exposed plates directly on the press cylinders, assuring correct registration from the very first copy. However, this approach permitted Pearl to use higher resolutions (up to 2,540 dpi), closer control over dot spacing and dot shape, and the expansion to larger press sizes without the penalty of longer setup times. In addition metal plates were introduced that achieved longer run lengths and true graphic arts quality levels.

The GTO-DI screen rulings are up to 175 lines/in. (6.9 lines/mm) in resolutions of up to 2,540 dpi. Imaging the plates is relatively fast — at 1,270 dpi, all four plates can be imaged in about 12 minutes. This is less time than it typically would take to image and process the same four pages on film in an imagesetter and substantially less than the average 40- to 50-minute offset-press makeready.

A disadvantage of the GTO-DI is its inability to run during platemaking. The Pearl upgrade has addressed this issue by reducing platemaking time from 20 min-

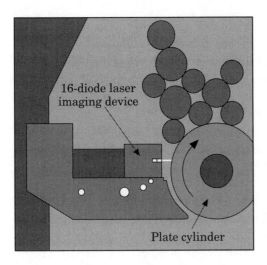 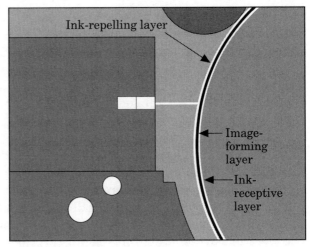

Basic principles of the Pearl imaging technology from Presstek, Inc.

utes to 12, and more recent RIPs reduce it further to about 8 minutes. During the 8-minute platemaking time, the Indigo E-Print 100 could print about 300 single-side color pages, and the Xeikon DCP-1 could print both sides of 300 sheets.

There is an upper limit on the number of impressions that makes sense with a GTO-DI. If the number of impressions exceeds 5000, it might be more efficient to print them on a press larger than the 14×20-in. (356×508-mm) size of the GTO-DI.

The GTO prints only four letter-size pages at once; a second pass through the press prints the reverse side, for a total of eight pages. Longer jobs would require gathering of signatures, so we would expect most GTO users to focus on jobs of eight or fewer pages, with perhaps the occasional 16-page job.

The question becomes what are the characteristics of an appropriate or ideal GTO-DI job:

- Run length: 500–5,000 copies
- Eight pages or fewer
- Available in PostScript format
- Process color

These are typical characteristics for marketing materials (brochures, product sheets, posters, and the like). Also in this category would be products such as book jackets, compact-disc (CD) inserts, some labels, and multicolored business forms. For this type of work, the GTO-DI is clearly better suited than either an electronic printer or a larger press.

Monochrome work (and spot-color work without critical registration) can be done at lower cost on a variety of other equipment. The typical GTO-DI will be used almost exclusively for process-color jobs.

## Quickmaster DI

At Drupa 95, Heidelberg and Presstek introduced a new generation of digital press, the Quickmaster DI. It is a four-color press with no option for an extra color or perfecting as was the case with the GTO-DI. The press is comprised of a single, quadruple-diameter (common) impression cylinder, around which were four color units. This unique system incorporates a common impression cylinder press design similar to the original Presstek design; simultaneous on-press plate imaging; waterless, nonphotographic plates; a highly advanced paper feed delivery system; and the latest on-press automation.

Each of these units has its own inking, plate, and blanket cylinders. It uses polyester waterless offset plates, which can be imaged at resolutions of 1,240 dpi and 2,540 dpi.

The plates are mounted automatically. A full roll of plate material can be loaded into each of the plate cylinders in around four to five minutes, and a full roll contains enough material for 35 plates. The Pearl technology used for imaging the plates is the same technology used with the GTO-DI. The Quickmaster can image at resolutions of 1,270 or 2,540 dpi, but because it uses a single 35-micron spot size from the laser diodes, there is no benefit to imaging at resolutions higher than 1,270 dpi. At that resolution, the plate is imaged in 6 minutes, compared with the 12 minutes on the GTO-DI. The press can run up to 10,000 impressions per hour.

Like the GTO-DI, the Quickmaster DI connects to a customer's prepress system for output of finished pages with no intermediate steps. As a result of simultaneous on-press plate imaging with the automatic plate register, plate handling, blanket

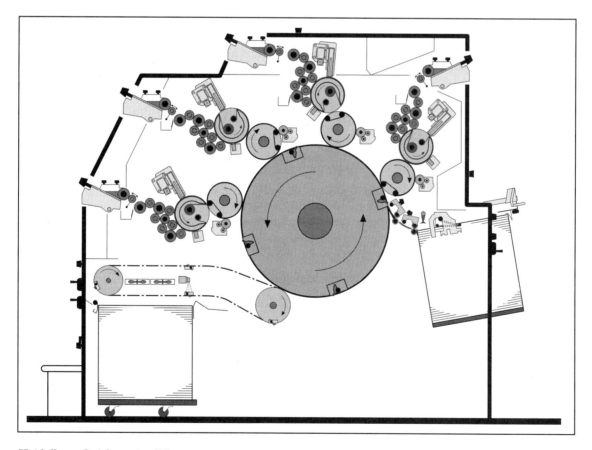

Heidelberg Quickmaster DI.

cleaning, and plate cleaning, the Quickmaster DI makeready times have been reduced to only 10–12 minutes. Combined with the prepress savings achieved in going direct to press, this results in cost savings of over 65% vs. traditional production with over 80% time savings.

The Quickmaster DI 46-4 has a built-in RIP server that is optimized for its type of work. The RIP runs on a DEC AlphaStation 200 4/233 computer. The PostScript interpreter, which comes from Harlequin, can preview rasterized data on the monitor before the platemaking.

The maximum image that can be printed is 13×17¾ in. (330×450 mm), and the maximum paper size is 13⅜×18⅛ in. (340×460 mm). This has been one of the most successful products ever introduced, with over 1,000 sold worldwide. Its current list price is $487,000.

### "Direct to Plate on Press" vs. Offset Presses

How does the DI technology fare in comparison with other small offset presses?

The difference between a normal GTO and the GTO-DI is the built-in platemaking facility. Naturally, this avoids the need for camera work and darkroom, stripping processes, and platemaking processes used in conventional offset. But all these steps could be avoided by using a PostScript platesetter of any kind, whether on the press or not. Presetting of inking levels (which the GTO-DI does automatically) can be done by scanning plates before they are mounted, so this advantage is not unique either. The waterless plates of the GTO-DI do not require a fountain solution, so ink/water balancing is not required. But this is a characteristic of any waterless plate.

So what are the unique advantages of the GTO-DI? It seems to us that there are two. First, it is difficult to obtain a waterless plate that can be imaged in a PostScript platemaker. Therefore, if you want to make plates digitally and run waterless plates, your only choice is the GTO-DI. With any other press, you will either have to make plates conventionally (from film) or deal with the ink/water balance issue on the press. That means either more consumables and labor (for the film and plate burning steps) or more adjustment (of ink and fountain solution) during job start-up.

Second, having the platemaking done directly on the press provides better register than conventional plate-mounting schemes. This lets the press start up in register, avoiding paper waste and saving some time at the beginning of the run.

### Omni-Adast DI

In 1996, Omni-Adast, a company with roots going back 600 years in the Czech Republic, introduced the first four-up on-press platemaking system. In quality and capability, it exceeds existing DI systems. There are four models, with a five-color perfector at the top of the line. (See table on the following page.)

The Omni-Adast dry offset presses use a 32-diode laser system that produces resolutions of 1016, 1270, 2032, or 2540 dpi. A new generation of plate material has run lengths capable of 50,000–100,000 impressions. Most importantly, all four or five plates can be imaged in under 7 minutes. This compares with 12 minutes on the GTO-DI and 6 minutes on the Quickmaster DI, both of which have 16-diode laser systems.

Omni-Adast says it has three orders in house right now and expects to have six or more systems installed by the end of the 1996, of which two will be five-unit presses.

**Omni-Adast DI Press Models**

| | 725PC | 725PHC | 745PC | 755PC |
|---|---|---|---|---|
| Units | 2 | 2 | 4 | 5 |
| Max. sheet size | 19¼₆×26 in. (485×660 mm) | 19¼₆×26 in. (485×660 mm) | 19¼₆×26 in. (485×660 mm) | 19¼₆×26 in. (485×660 mm) |
| Speed | 10,000 iph | 10,000 iph | 10,000 iph | 10,000 iph |
| Pile | Low | High | High | High |
| Price | $540,000 | $550,000 | $895,000 | $1,050,000 |
| Other | | | | Chill system Shortwave dryer Perfecting |

In terms of cost, the four-color version compares favorably with the two Heidelberg DIs and very favorably with a four-color and a two-color GTO using traditional approaches.

In the following table, the first two devices are traditional GTO presses running wet offset, and the last three are DI presses. Paper is not included in these numbers. Based on our cost model, the four-color version of the Omni-Adast at $895,000 comes in cheaper than the GTO-DI at $603,400. That is because of the number of pages being printed at one time. We have plugged in a higher cost for the plates (they are larger) and also factored in the faster platemaking time based on the 32-diode laser system. Because hybrid presses have quicker makeready and run very fast with a

**Comparative Costs per Unit for Various DI Presses and GTO Models**

*8.5×11 in., Four Color, One Side*

| Quantity | GTO 4C | GTO 2C | GTO-DI | QM-DI | O-A 5C |
|---|---|---|---|---|---|
| 50 | $3.33 | $4.46 | $1.04 | $0.86 | $0.98 |
| 100 | 1.67 | 2.23 | 0.52 | 0.43 | 0.49 |
| 500 | 0.34 | 0.45 | 0.11 | 0.09 | 0.10 |
| 1000 | 0.17 | 0.23 | 0.06 | 0.05 | 0.06 |
| 5000 | 0.04 | 0.05 | 0.02 | 0.01 | 0.02 |

*8.5×11, Four Color, Two Sides*

| Quantity | GTO 4C | GTO 2C | GTO-DI | QM-DI | O-A 5C |
|---|---|---|---|---|---|
| 50 | $6.48 | $8.93 | $1.95 | $1.75 | $1.92 |
| 100 | 3.25 | 4.48 | 0.99 | 0.89 | 0.98 |
| 500 | 0.70 | 0.95 | 0.23 | 0.21 | 0.23 |
| 1000 | 0.35 | 0.48 | 0.13 | 0.12 | 0.13 |

larger sheet of paper containing multiple pages, they have the best cost per copy numbers for print reproduction. Thus all the DI machines come to $0.12–0.14 per four-over-four copy at a run length of 1,000.

One of the ideas that came out of the EPEX 1996 press conference announcing the Omni-Adast DI presses was an approach to personalization. For printing that requires the imprinting of dealer addresses or other custom information on a four-color sheet using the black plate, the first job could be imaged with all four plates. For the second job, the cyan, magenta, and yellow plates would remain and a new blank plate would replace the black plate. It would then be imaged with all the black data plus the custom data. Since everything is in register, you are all set.

## Other Hybrid Presses

At Drupa 95, MAN-Roland showed two fascinating new prototype press systems, one for offset printing and one for gravure, both with the name DICOweb. (Chapter 33, "Futures," discusses these devices in more detail.)

The DICOweb Offset was based on an erasable plate material where the image was created by thermal transfer. The press ran as a conventional offset press with conventional ink and fountain solution. The image is created on the plate by thermal transfer. A ribbon of hydrophobic material is held in contact with the plate while a laser images it from the back. Where the laser heats it, material is transferred to a hydrophilic plate surface. Ink will stick to the imaged plate only where the thermal transfer has occurred.

The prototype machine uses a 64-channel imaging system developed by Creo. The resolution is 2,400 dpi. The imaging time is the equivalent of about 5 minutes for an A3 (11.7×16.5 in.) image, but this can be further reduced, according to MAN-Roland, to 2 minutes.

The press runs as a normal offset press, with conventional inks and fountain solution. The press speed is about 2 m/sec. It can be used for runs from 500 to 5,000 copies, although longer runs are possible. After the print run, the image is dissolved from the plate by a special solution and the plate is ready to be imaged again for another job. The plates can last for at least twenty jobs, after which they are placed by new ones.

The DICOweb Gravure comprises a system for on-press image engraving on a reusable cylinder. The cylinder (actually just a metal sleeve) is supplied pre-engraved with full depth cells and then, when preparing for a job, is coated with a dark, waxy polymer. This fills all the cells. Then the polymer is selectively removed by evaporation. The amount of laser exposure controls the depth of the cells. The sleeve is reusable and stays on the press for at least twenty jobs and is designed for print runs from a few hundred up to 5,000–10,000 copies. When the print run is done, the remaining polymer is removed and the process starts again. All the steps involved in coating, exposing, and cleaning the cylinder are automated, and they take just five minutes in the prototype system.

Another interesting characteristic of the system is that the image resolution is not limited to the resolution of the gravure cells. For fine text and line art, areas of polymer within a single cell can be selectively cut away, eliminating "jaggies" typical of publication gravure.

## Zap the Gap

Vibration is the printing press killer. Web presses could operate at higher speeds if they could rotate their cylinders with less vibration. The main culprit is the plate cylinder because of the gap where the leading and trailing edges of the plate meet.

With new on-press platemaking technology, the cylinder could contain a plate that was 360° around. The cylinder itself could be re-imageable many times or a replaceable sleeve could do the job.

Presstek developed a 360° plate for MAN Roland a few years ago, but we have not seen it in a product yet.

## Image Carriers

The essence of reproduction technology is the image carrier. Even purely digital printers have an image carrier. It is usually an organic photoconductor (OPC) of some sort that can hold an electrical charge. Images are built up as lots of charges, and toner in some form is attracted and later transferred to paper. The OPC is then cleaned and re-charged for the next image.

Printing presses use a non-reimageable image carrier. It can be produced offpress or, as Presstek proved, onpress. Only ink jet and some esoteric imaging technologies that utilize special papers bypass the image carrier.

Imaging an image carrier on the press makes sense. It short circuits the digital workflow, which has, in less than a decade, gone from computer-to-film, to computer-to-imposed film, to computer-to-plate, and now computer-to-onpress-image carrier, and perhaps to press cylinder.

As previously mentioned, MAN Roland has demonstrated the DICOweb offset technology that uses its own re-imageable image carrier and an exposure system from Creo. A lot of what happens will depend on Presstek. Presstek's main patent covers the production of a plate on a printing press. Is a re-imageable cylinder or sleeve a plate? Legions of lawyers will argue both sides. Plates have always been imaged one time. Are they still plates if they are imaged multiple times?

One way or another, onpress production of image carriers will happen. The first benefit will be cutting the time and cost of workflow; the second benefit will be the re-engineering of presses with adapted or new technology to run faster and more efficiently because the gap is gone.

By 1997 we will see the first of computer-to-press cylinder, and real products should hit the market by 2000. The future is just around the corner.

## Conclusions

The concept of imaging a plate directly on press is indeed clever: there is no need for worrying about registration, and the ink profile is automatically generated. Moreover, no additional room (and thus additional rent) for placing the imaging recording devices is required. For example, in the case of purchasing a platesetter, the room where it will be placed must be taken into consideration.

Presstek was the first to implement the concept of imaging plates directly on press, while successfully creating a new market for their product. The most significant factor of Presstek's invention is not technological but economic by making short-run, good-quality offset printing affordable. The latest model, the Heidelberg

Quickmaster DI, takes a further step toward short-run digital printing that can compete, head to head, with other exclusively digital presses. Above all, Presstek's technology makes use of the lithographic printing process that printers are familiar with and can trust not only for its reliability but also for its high quality too.

# Chapter 27

# Xeikon DCP/32D

Like other digital press manufacturers (i.e., Indigo), Xeikon is a relatively new company. Based in Mortsel, Belgium, it was founded in 1988. The initial funding came from AGIF (Agfa Gevaert Investment Fund). Since then at least six other companies have invested in Xeikon, but Agfa remains the single largest shareholder with approximately 25% of the shares.

Agfa's investment is no great surprise because they helped pioneer the underlying technology. In the early 1980s Agfa marketed the first 400-dpi laser printer with the P400. It was the first laser printer with resolution greater than 300 dpi and was rated at 18 ppm, considered extraordinary at that time.

The P400 was targeted for a new market called the emerging workgroup market. The person in charge of this technology group at Agfa was Lucien De Schamphelaere, who later ran the AGIF venture funding group. However, contrary to popular belief, there are only a few senior managers at Xeikon who came from Agfa. The balance of the staff came from other high-tech companies in Belgium.

Today, De Schamphelaere is chairman and president of Xeikon. The company was founded in 1988 by Lucien De Schamphelaere. The DCP-1 was announced at a press conference on June 21, 1993. The Agfa Chromapress was shown in September 1993 at the IPEX exhibition in Birmingham, England. The first production units were shipped in April 1994.

## Financial Update

As we go to press, Xeikon N.V. (NASDAQ: XEIKY) announced its results for the first quarter of fiscal 1996. For the quarter ended March 31, 1996, revenues were $18.8 million as compared to $19.9 million in the year earlier period. Gross profit for the quarter increased 62% to $4.2 million from $2.6 million in the first quarter of 1995, reflecting an increase in gross margin to 22% from 13%. Net income for the quarter increased to $275,000, or $0.01 per share, from a loss of $52,000 in the same period during the prior year. Earnings per share for the first quarter of 1995 have not been presented as the company was private until March 19, 1996.

Xeikon sold 66 DCP-1 systems during the quarter, bringing the total number of systems shipped since the start of commercial production in mid-1994 to 450. Alfons Buts, the company's chief operating officer, noted, "Our installed base of regular users continues to grow significantly — a reflection of steady demand — and provides the platform for future sales of consumables such as toners and developers." He pointed out that while Xeikon shipped approximately the same number of systems in the first quarter of 1995, a significant number of those systems were for use in systems integration, testing, and training at Xeikon's OEM partners. Systems sold today are for regular end-use by commercial printers, in-house print shops, etc.

He noted that the lower level of revenues for the quarter reflected a significant inventory buildup of consumables in the distribution channels in the first quarter of 1995, as well as a new supply agreement entered into with Agfa-Gevaert in January 1996. Under this new agreement, Agfa-Gevaert is now selling the consumables, which it manufactures for Xeikon, directly to its customers, paying the relevant commissions to Xeikon. The arrangement has the effect of lowering Xeikon's total revenues because it eliminates the resale of consumables by Xeikon to Agfa-Gevaert, but the commissions are included in revenues and subsequently in gross profit. Xeikon raised net proceeds of approximately $65 million through its initial public offering on March 19, 1996.

## Distribution Channels

Despite the financial support of Agfa, Xeikon is an independent company. The DCP-1 is manufactured in Mortsel and, until recently, was sold through two channels in the U.S. — AM selling the Xeikon unit with several different front ends, and Agfa adding its own front end and resells the unit as the Chromapress. Although not a formal distribution channel, a few large printing companies (e.g., Donnelley and Moore) have

The Xeikon DCP-1.

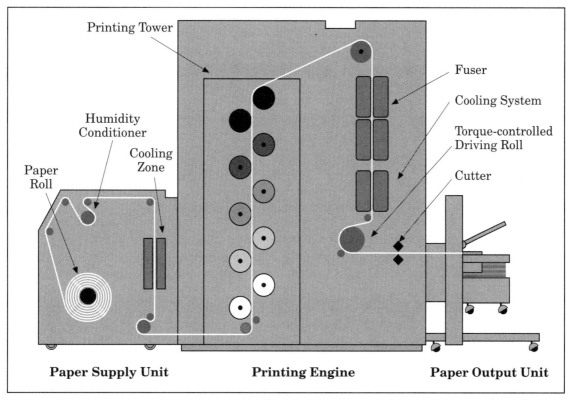

The general system schematic is the same for all engines based on Xeikon, including the Agfa Chroma-press.

made commitments to provide on-demand printing services and have bought a fair number of units. These printing plants are beta test sites and have performed their own research and development to customize the equipment to their own specifications. As of January 1997, AM is no longer a distributor of the Xeikon system.

## Electrophotographic Process

The electrophotographic process is a well-established process used in a variety of laser printing and copying devices. The four basic steps of the process are described.

In the print tower, the engine is made of ten print units, eight of which are used to print four color images on both the front and back of each side of the web. The building block of the technology is the organic photoconductor (OPC) drum, which consists of an aluminum core covered with a light-sensitive material. In the first step in the process, the drum is charged.

If you have ever changed a laser printer or copy machine cartridge, you probably know about the corona. The corona, a wire within a metal enclosure, sits very close to the drum and is attached to a high-voltage power supply. When the corona is charged, the air becomes charged with ions and/or electrons. These particles charge the drum surface. Another device called a scorotron, a special kind of corona, is responsible for controlling the electrostatic charge.

In the second step, a light-emitting diode (LED) exposes the drum to precise levels of light, generally from a laser beam. On the Xeikon, a fixed bank of 7424 LEDs expose the drum to varying amounts of light. When the light hits the drum the OPC layer becomes conductive, and the exposed areas lose their charge. Precisely controlled amounts of light from the LED array build the image.

In the third step, the toner particles are magnetically attracted to the drum. As the page passes the toner cartridge, the toner and the magnetized carrier particles from a magnetic brush pass down the drum, transferring toner particles that adhere to the latent electrostatic image on the drum.

In the fourth and final step, the image is transferred to the paper. The developed image is transferred to the web of paper, then bonded to the paper with heat.

## DCP-1 Specifications

The first product from Xeikon is the DCP-1, a web based device which uses copier or electrophotographic technology. The unit is housed within three linked cabinets that Frank Romano often compares to the metal engineering of a battleship. The Xeikon engine is contained in the first two cabinets. The first cabinet contains the paper supply. The second and largest cabinet contains the imaging units, which are stacked one on top of another, and the fuser.

In the second cabinet, the rollers are imaged, and the toner comes in contact with the paper. The web of paper is transported between eight color rollers, two for each color. It has eight identical printing units, five on each side of the paper, corresponding to the cyan, magenta, yellow, and black units of a conventional four-color press. The fifth allows for future expansion. The fusing of the toner to the paper occurs in the second cabinet. The DCP-1 uses a noncontact fusing system. The advantage of this system is the wide range of papers that can be used. The press is rated for paper weights from 80 to 200 gsm, which translates to 54- to 135-lb. stock.

Unlike some traditional web presses, the DCP-1 does not support changing of paper rolls on the run, sometimes called the "flying paster." The maximum paper width is 12.6 in. (320 mm), which is well suited to jobs made up of 11×17-in. or A3 size pages.

Depending on the paper stock, the roll can range between 2,000 and 4,000 ft. (610–1220 m) in length. With 100-gsm paper, one roll allows for over two hours of printing.

After imaging, the paper is passed into the third unit, which contains a built-in sheeter and stacker assembly. The sheeter cuts the paper and deposits it into a removable output bin. Also included are an automatic job separator and a special tray for unprinted and test sheets. An optional in-line finisher allows sheets to be collated and folded into booklets.

In terms of speed, it prints 35 letter-sized sheets/minute, simplex or duplex. Because the press images both sides of the web simultaneously, there is no speed sacrifice for duplex printing, meaning the Xeikon can print 4200 letter-sized pages an hour.

Unlike sheetfed devices, which are restricted in size in two dimensions, the DCP-1 is only restricted in one — the width of the roll. The length was originally limited to A3 sheets, but subsequent software releases expanded the size. Because the web speed is constant, a longer sheet length takes longer to produce.

Using copier-based technology after imaging, the toner is applied and permanently fixed to the paper by a noncontact, hot fusing system. Like other digital presses, the consumables are only available from the manufacturer. The main consumables in this case is the toner.

The system is capable of printing a fifth color, but this capability has not been implemented at this time. The registration system is based on Xeikon proprietary technology and is so good that many users claim they do not need to trap their files.

Like the Indigo, no plates are used in imaging, but unlike the Indigo, there is no ink to adjust. The cabinet only needs to be opened for maintenance or to rethread paper.

## Resolution

Resolution for copier-based technologies is different than traditional four-color process. In traditional four-color printing, film is created with a resolution of approximately 1000 or 2000 dpi. The dots are fixed in size but "clumped" together to create different size "spots." In this system, the maximum levels of gray depend on the resolution and line screen.

The Xeikon prints at 600 dpi for color, but incorporates a variable spot function that allows each 600-dpi spot to print with any of 64 levels of gray by applying different amounts of toner. Spatial resolution of the DCP-1 is 600 dpi for each color. But unlike an imagesetter, the system has a variable spot intensity; each 600-dpi spot can have 64 levels of gray (LOG) by applying different amounts of toner. As a result, the image quality is more like a 2,000-dpi imagesetter than a 600-dpi laser printer.

Andrew Tribute (*Seybold Reports*, June 28, 1993) has compared the DCP's resolution to gravure printing. He says that the resolution in gravure printing is also approximately 300 dpi with multiple gray levels. In gravure printing, anti-aliasing can be used to improve the appearance of linework and type. Taken to the end of this argument, the DCP's 600-dpi resolution and 64 gray levels, therefore, could theoretically approach gravure in quality. Xeikon is not now using any anti-aliasing, but Xeikon's OEM customers or large clients may in the future.

Screening like the resolution and LOG mentioned earlier are unique for the digital presses. Again Andrew Tribute has an interesting perspective when he says that the Xeikon technology "is similar to the now-defunct LaserGravure system that Crosfield worked on many years ago." There, the laser beam cut a groove into the cylinder, and the width of the groove controlled the amount of ink color that was printed. With Xeikon's DCP-1, the screening appears to be made up of four lines of continuously varying width, each at a different angle. Under a loupe, some areas look like small diamond-shaped cells, while others look like a number of intersecting lines. However, none of this is obvious to the unaided eye. Xeikon also offers traditional screening with a clear-center rosette at 170 lpi on the DCP-1/F.

## LED Imaging

Unlike many black-and-white on-demand systems that use laser beam technology to create the pages, the Xeikon uses light-emitting-diode (LED) imaging technology.

According to Andrew Tribute (June 28, 1993 ) of the Seybold organization, "Light-emitting diodes, like laser diodes, generate light by passing a current through a silicon crystal that has been 'doped' with selected impurities. Unlike lasers, though, the light from an LED is not coherent, and is thus harder to collimate and focus. Rather than sweeping a single light beam across the image from a distance, LED printers use arrays of tiny LEDs placed very close to the image. Each LED is responsible for forming a single dot of the output."

The DCP-1 uses standard LED wafers, assembled into arrays of 7,400 diodes, spaced at 600 per inch. Xeikon's printing process requires that each diode provide a variable amount of output. Xeikon also uses a proprietary method to normalize individual LED exposure to ensure consistent image density across the LED array.

## Controller

The standard Xeikon print engine is driven by PostScript code. As of this writing it uses Harlequin's Level 2 interpreter running on a 66-MHz Pentium PC, supplemented by Xeikon's EISA-bus screening card. The controller creates four bitmaps, one for each color. Output from the RIP is buffered to disk. Bitmaps are stored on up to 12 GB of disk space and then transferred to the page buffers. Each color printing unit has its own 72-MB image-buffer memory. An ASIC with a maximum data transfer rate of 192 Mbits/second feeds data from memory to the LED imaging array.

All of the separate processes in the machine are handled by a set of distributed microcontrollers that are connected to a supervising controller by means of an optical network.

The GUI (graphical user interface) is a Microsoft Windows interface. (The RIP also runs with a Windows user interface.) The interface into the system is via Ethernet using Ethernet protocols. No input buffering of data is provided in the system, and it is strongly recommended that an OPI image server be used.

## Maintenance

According to Xeikon, the system is designed for continuous operation with only small amounts of maintenance. Scheduled operator maintenance is approximately 30 minutes per day with an additional hour of work once each week.

Xeikon recommends preventive maintenance between 100,000–150,000 A4 sheets (or 300,000 duplexed impressions) or 70 hours of continuous printing. This is conservative, according to Xeikon, and projects that the maintenance time periods between scheduled maintenance will become longer as the technology improves.

Compared to the daily and weekly maintenance of a conventional press, the Xeikon schedule is small. However, if you calculate maintenance based on number of impressions, the traditional press fairs better.

## Xeikon Drupa Announcements

- **MultiPage support,** which allows Xeikon users to switch jobs "on the fly." The press produces electronically collated multipage documents up to 64 pages in length. QuickProofing prints proofs in the midst of production runs without stopping the press.
- **Xeikon Variable Data System,** which allows text, line-art, and images to be varied on every consecutive page in black or in full color without affecting the printing quality or speed.
- **Barco Graphics PrintStreamer,** a page buffer, co-developed by Xeikon, that stores 500 magazine pages (5,000 text pages) and allows production to fully collated color documents and books.
- **Non-PostScript input formats.** Barco Graphic's dual-channel FastRIP/X provides the Xeikon press with an excellent gateway to non-PostScript prepress platforms including Scitex- and Hell-native formats.

- **In-line finishing.** A modular in-line finisher produces stitched, folded, and trimmed booklets. Other features include cover sheet insertion, stacker, test tray, and bypass.

## New Features

Xeikon announced the introduction of the Xeikon DCP/32D, its second-generation digital short-run color printing system. The system includes new features in hardware, software, and consumables that combine to improve quality and productivity and to lower costs. Xeikon will offer an upgrade package to users of its first-generation system, the DCP-1, which was commercially launched in mid-1994. Full production of the DCP/32D is already under way, and the first units are expected to be shipped in mid-June.

The Xeikon DCP/32D uses an enhanced imaging component, called the GEM module, to print images with glossy finish and superb color saturation equivalent to offset printing. It also features second-generation software and consumables, such as toners and developers, which combine to lower production and maintenance costs and make the press much easier to use. The new system will also feature Xeikon's One-Pass Duplex Color technology, which enables its systems to print full color simultaneously on both sides of the paper. The Xeikon DCP/32D will be able to process nearly twice as many jobs as any other commercial digital color press and significantly more short-run jobs than the most advanced offset presses in a given amount of time.

**DCP/32S for Single-Sided Printing.** Xeikon, which re-named its product based on the paper width handled (32 cm) and the number of sides printed at one time, has also introduced a single-sided version. Based on technology and toners recently announced in the DCP/32D (D for duplex), the new DCP/32S is a single-sided printing system targeted at pressure-sensitive and other adhesive-backed label printing applications.

According to Xeikon, the label printing industry has worldwide revenues of $15 billion (Indigo says $14.8 billion worldwide and about $6 billion U.S.), so it is no wonder that both firms are eager to enter this lucrative market. The debut of the new digital press was at the Label Expo show in Chicago in September 1996.

In addition, Xeikon also announced an OEM agreement with Nilpeter for worldwide distribution of the DCP/32S. Nilpeter A/S, headquartered in Denmark, has service facilities in sixteen countries. The company designs and manufactures rotary label presses for letterpress, flexo, screen, and hot-foil processes. Nilpeter will combine the DCP/32S with a range of finishing and material handling options to provide a complete label printing system.

Almost 82% of all labels are printed in pressruns of less than 100,000, and 35% have personalization. In addition, 80% of all labels are roll-fed. Nilpeter, one of Xeikon's partners in the label market, is developing a reputation as a company intent on pursuing advanced technology. The three largest markets for labels are health and beauty, pharmaceutical, and horticultural. There are 573 label converters/printers in the United States, most using flexographic presses that are 9, 11, and 13 in. (229, 279, and 330 mm) wide.

**Xeikon RIPs.** The following table lists the initial RIP offerings — six in all — from Xeikon via its dealers. Agfa uses its own RIP technology with the Chromapress.

| Press | Raster Image Processor |
|-------|------------------------|
| DCP-1 | Pentium-based Harlequin ScriptWorks RIP |
| DCP-1/S | DEC Alpha Harlequin ScriptWorks RIP |
| DCP-1/F | DEC Alpha PostScript-only Barco Graphics FastRIP/X |
| DCP-1/F2 | DEC Alpha Dual channel PostScript & CEPS Barco Graphics FastRIP/A |
| DCP-1/C | DEC Alpha PostScript-only Barco Graphics FastRIP/X with PrintStreamer for variable data printing |
| DCP-1/C2 | DEC Alpha Dual channel PostScript & CEPS Barco Graphics FastRIP/A with PrintStreamer for variable data printing |

Xeikon has announced the DCP-1/S RIP, which uses the DEC Alpha RISC-processor and the Harlequin ScriptWorks PostScript Level 2 RIP.

**New RIP Configurations.** With the new DCP/32D engine, Xeikon can now configure up to six different RIP options, instead of the previous four. These options are:

- **DIS/B.** A 60-MHz Pentium running the Harlequin's ScriptWorks PostScript interpreter software, the entry-level configuration. (We wonder why anyone today would even consider buying a 60-MHz Pentium, when 90-MHz and 120-MHz units can be obtained for a very small extra cost. Since this is the only Pentium RIP offered by Xeikon, it's obvious that the company is much more interested in selling the other five configurations using DEC Alpha AXP processors.)
- **DIS/S.** A 233-MHz DEC Alpha AXP running ScriptWorks.
- **DIS/F.** A 233-MHz DEC Alpha AXP running Barco Graphics' FastRIP/X Post-Script interpreter. This is currently at Level 1 PostScript. Barco will upgrade the FastRIP to Adobe's Supra architecture, but we have no indication now if this will be available as an upgrade to the existing FastRIP.
- **DIS/FP16.** A 233-MHz DEC Alpha AXP running Barco's FastRIP/X with an integrated Barco PrintStreamer for extended page image memory. It provides 16 GB, using multiple RAID disk drives, for handling large documents and for variable data printing.
- **DIS/FP32.** A 233-MHz DEC Alpha AXP Barco's FastRIP/X with an integrated PrintStreamer and 32 GB.
- **DIS/SP32.** A 233-MHz DEC Alpha AXP running ScriptWorks, plus an integrated PrintStreamer with 32 GB of extended page-image memory.

**BARCO Graphics Productivity Features for PrintStreamer.** At Seybold San Francisco 1996, BARCO Graphics introduced some new features to PrintStreamer, Barco's large and extremely fast buffer for storing ripped pages between the BARCO Graphics FastRIP/X and the Xeikon Digital Color Press (DCP/32D). PrintStreamer enables digital presses to be used for sophisticated applications by allowing collated printing of variable information. Concurrent input and output streams permit the

PrintStreamer to simultaneously accept data coming from the FastRIP/X and send previously stored data to the press, improving its capacity utilization.

In addition to these features, BARCO Graphics has now released backup and restore software for the PrintStreamer. This software allows ripped data to be backed up on any medium that can be driven by a SCSI device (RAID disks, magneto-optical disks, tape, etc.). Once backed up, the data can be restored on the PrintStreamer or transferred to another PrintStreamer. The key benefits of the new software include the following:

- Possibility for off-line ripping of digital printing jobs
- The use of multiple RIPs for one press
- Fast reprints of backup jobs
- Storage extension for the PrintStreamer

**Larger Reels of Paper.** Roll Systems, USA has an unwinder unit that lets rolls of paper with 50-in. (1270-mm) diameters link to the Xeikon in addition to the present 16-in. (406-mm) rolls. This unit had been shown at Drupa in the IBM booth. The benefit with larger-capacity rolls is simply less time between roll changes. Like imagesetting film, the Xeikon paper leader and trailer cannot be imaged because they have lost the conditioning for holding an electrical charge. The unwinder cuts paper waste by cutting re-loading times. This assumes that the user remains with the same paper type through a number of jobs.

**Instant Job Switching Now Standard.** An instant job switching feature will be standard on all Xeikon systems, including entry-level versions, and will be provided to all existing users as a free upgrade. It allows one job to follow another job without stopping the print engine. All jobs that are in the print queue will be printed nonstop.

**Printing on Polyester Film.** The Xeikon system can now print on polyester films from Swiss companies Messerli and Folex and Israeli-based Hanita. Polyester film has applications for architectural plans, maps that are handled often, certificates, etc. The material is usually difficult to handle on sheetfed presses.

**Private-I for Personalization.** This is a new software package for the creation of variable data documents — personalized printing of text, line art, or images. The Xeikon Variable Data System must be installed.

Production of personalized products is a multi-step process. In the design step, the basic layout and content of the document is made, using page layout programs that can produce EPS files. The variable content of the document is then determined. It could be in any database that can output DBF format, which is virtually standard with every database program.

In the next step, the variable information is merged with the basic document design. Private-I facilitates this step, claiming that it assists graphic designers.

There are two image memories in the Xeikon Variable Data System: one to hold the fixed data (the page layout and master design) and the other to contain the variable field information (text, line art, or images). As the job is processed, the two types of files are merged on the fly by the print engine.

**Xeikon DCP/32D Specifications.** The following are specifications for the imaging engine of the DCP/32D:

- High-quality, short-run CMYK digital imaging based on electrophotography
- 4,200 four-color 8.5×11-in. (216×279-mm) impressions/hour (2,100 duplex sheets)
- 2,100 four-color 11×17-in. (279×432-mm) impressions/hour (1,050 duplex sheets)
- 600-dpi spatial resolution
- PostScript Level 2 RIP
- Webfed roll input
- Maximum image area: 12.1×408 in. (307×10,360 mm)
- Paper width up to 12.6 in. (320 mm)
- Paper weight up to 12 pt stock.
- Variable cut-sheet length, user-programmable

The following are specifications of the PostScript RIPs for the DCP/32D:
- Open-ended architecture is compatible with all standard DTP platforms and PostScript software
- RIP output spooling allows pre-processing and continuous printing
- Compatible with Ethernet
- Harlequin software RIP, PostScript Level 2 compatible, running on DEC Alpha chip (233 Mhz) with proprietary Xeikon screening at 150 lpi or
- FastRIP software RIP on a DEC-Alpha (233-MHz) chip printing clear-centered rosette pattern at 170 lpi and optional automated trapping and imposition; color match to SWOP or Euroscale standards

Xeikon has begun manufacturing and shipping what it calls a "second-generation" digital color printing system as a replacement for its DCP-1. Called the DCP/32D, the new system includes:
- A series of refinements to the original print engine technology
- A new software release
- For Xeikon-brand systems, a broader selection of RIPs to drive the print engine

All of the announced improvements can be added to current DCP-1 customer sites as field upgrades. Agfa, one of Xeikon's OEMs and the manufacturer of its consumables (toner and developer), has already announced the improvements (except for the Xeikon-specific RIP options) as enhancements and upgrades to an improved version of its Chromapress.

In addition to the fanfare surrounding those developments, Xeikon more quietly has made some evolutionary changes to the engine (and the configurations in which it sells it) since initial shipments began. However, the basic engine implementation, using its one-pass duplex technology, remains unchanged, offering a printing speed of 2,100 duplex, full-color pages per hour.

**The Glossy Look of Offset.** One of the most significant of the enhancements to the print engine is a new toner-fusing mechanism intended to give the output a higher gloss and closer resemblance to offset printing. Xeikon calls this optional enhancement Gem; Agfa calls it OmniGloss. It consists of a new, two-part fuser and finishing system as a replacement for the original toner fuser in the engine. With the new capability, the output medium is heated and toner is partially fused at the first station; the medium then passes to the second fusing station of pressurized rollers, which results in a calendering effect on the media and a glossier appearance from the fused toner. A new, heavier-duty web puller motor is also part of the option.

**Broader Range of Stocks.** As a byproduct of the two-part fusing process, the fusing temperature in the engine can be lower, thus allowing the use of a broader range of stocks, including coated and heavier ones. The new fusing technique also makes the toner less susceptible to scratching.

The stronger web puller motor helps accommodate heavier stocks, so the machine is better able to print covers, business cards, display materials, and the like. The engine can now accommodate stocks from 40 to 170 lb. (60–250 gsm). It is possible to tune or turn off the second fusing station under operator control.

**More Cost-Effective Consumables.** Another key enhancement to the print engine is a set of reengineered, more cost-effective consumables — both toner and developer — developed and supplied by Agfa. (The developer is the component that carries the toner particles and distributes them to the electrostatically charged areas of the printing surfaces.)

Agfa is the sole manufacturer of consumables for the Xeikon engine, which is private-labeled for IBM and AM, although there has been some discussion, we understand, about developing second sources. We know that IBM, for example, has considerable experience in this area.

These "second-generation" microtoners and developers enable lower toner consumption by up to 15–20% per page, Xeikon says, a longer developer life of more than 150,000 sheets (up to three times the life of the previous developer); and enhanced toner coverage. This last factor, coupled with the ability to run more media types, greatly increases in number the kinds of jobs that can be printed on the press. Agfa states that the enhanced consumables reduce the variable costs associated with printing by one-third, which undoubtedly will have an effect on the economic model and ROI for the devices. The new toner and developer will be available around the beginning of the third quarter.

**Longer Drum Life.** Related to the lower cost of ownership are new preventive-maintenance tools and procedures for the upgraded engine, which result in substantially longer lives for parts and other consumables. For example, the OPC (organic photoconductor) drum will be able to achieve a useful life of 200,000–300,000 sheets. Likewise, the longer developer life means less preventive maintenance from fewer developer changes (a change takes about 4–5 hours), thus less downtime.

**New Software.** Beyond these refinements, a new release of software for the new Xeikon models provides such improvements as a friendlier operator interface, more automated process controls, and improved conditioning of the paper and print stations at the engine, which are intended to provide predictable performance and shorter time in startup. It is also possible for a key operator to perform extended maintenance.

All of these enhancements contribute to making the new DCP/32D (and others based on the same engine) more suitable for heavy-duty use and monthly volumes of up to 500,000 pages.

**System Pricing.** DCP/32D prices vary according to the RIP option chosen. The base level with the DIS/B RIP costs in the region of $280,000. A high-end configuration using either the DIS/FP32 or DIS/SP32 RIP is priced in the $450,000 range. All current DCP-1 users can upgrade their presses in the field to DCP/32D specifications.

## Xeikon and Xerox Settle Patent Dispute

Xeikon and Xerox Corporation announced jointly the binding settlement of a pending patent dispute and infringement suit filed by Xerox. Xerox had filed the suit on March 27, 1996 with U.S. Federal District Court at Rochester, alleging that Xeikon had violated certain Xerox patents. The settlement also defines a set of procedures to discuss any future patent disputes.

The companies have also entered into an agreement that defines the terms and conditions for a possible original equipment manufacturer (OEM) arrangement to apply should Xerox elect to distribute Xeikon's digital color printing system. Under this agreement, Xerox will purchase from Xeikon a small number of Xeikon systems for evaluation and testing.

## Barco Relationship

Xeikon N.V and BARCO Graphics NV reported that they reached an agreement whereby:
- Xeikon will supply as an OEM the BARCO Graphics PrintStreamer configuration through its own international network of distributors.
- BARCO Graphics will sell the DCP-1/F as an integral part of the total BARCO Graphics solution for digital printing.

Xeikon's value-added distributors (VADs) will sell the BARCO Graphics 32-GB and 16-GB versions of the PrintStreamer. The PrintStreamer is a spooler device that enables collated printing of pages for book production and also enables text and picture databases to be merged into the color page providing unique variable data functionality. It can store approximately 1,000 typical magazine pages, or 10,000 text pages.

As an OEM for the Barco Graphics PrintStreamer, Xeikon is able to incorporate this technology directly into its DCP-1 presses and offer existing users the opportunity to upgrade their existing DCP-1 units. This high level of functionality allows Xeikon users to explore new market-capturing applications and means Xeikon customers are able to capitalize on the most important features of digital printing, namely, full-color variable-data printing and providing a collated document from the press.

Units sold by BARCO Graphics will be delivered, installed, and serviced by the local Xeikon distributor. Training and additional consumables will also be provided by the Xeikon distributor.

# Chapter 28

# Agfa Chromapress

The Agfa-Gevaert Group is a leading worldwide manufacturer of imaging products and systems, with annual worldwide sales of $4.5 billion. Agfa-Gevaert manufactures and markets products and systems for the electronic and photographic prepress (Agfa Graphic Systems), medical and technical diagnostic (Agfa Technical Imaging Systems), and amateur and professional photographic markets (Agfa Photo Imaging Systems).

Headquartered in Mortsel (outside Antwerp), Belgium, Agfa-Gevaert is wholly owned by Bayer AG. In the United States, Agfa operates as a division of Bayer Corporation, with headquarters in Ridgefield Park, N.J.

The Graphic Systems Business Unit is responsible for the Chromapress as well as scanners, color management software, fonts, servers, imagesetters, film recorders and consumables. Chromapress was developed in Belgium.

Chromapress combines the Xeikon print engine with Agfa's front-end expertise. Agfa has a wealth of prepress experience from both the merger with Compugraphic and electronic hardware and software developments in Belgium. Agfa is also a leader in electrophotographic systems due to its copying group in Europe.

Agfa announced Chromapress in September 1993. It is an integrated computer-to-paper system for on-demand, affordable, high-quality color printing. Chromapress is said to be a complete solution, incorporating prepress through reprographic technologies to support the production of timely, cost-effective color documents. This "systems" philosophy extends from the creative concept to PostScript files and on into printed and finished documents, and it embraces the critical ownership issues of training, service, and long-term support.

The Chromapress system is a "turn-key solution," which means that Agfa sells a package including training, support, and most of the consumables. It is one of a few systems — if not the only one — based entirely on Macintosh-based software. The Chromapress software is designed in a workflow metaphor with divisions for job tickets, job scheduling, input queuing, color management, imposition, multiprocessor ripping, and automatic duplexed printing.

The core technologies of the Chromapress are evolutionary. The print output engine, for example, is electrophotographic or xerographic. Some team members

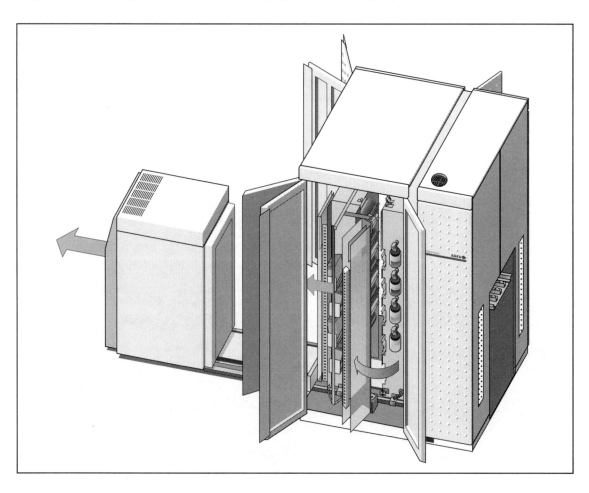

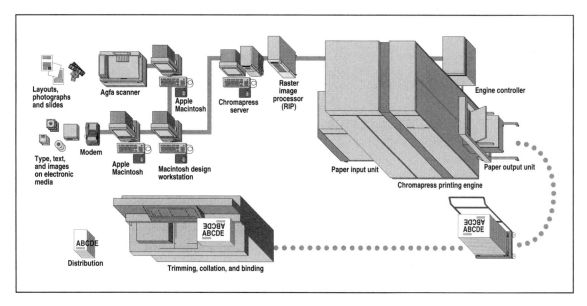

worked on Agfa's first laser printer — the P400, which at its introduction was 400 dpi and the fastest PostScript printer available. Chromapress also incorporates color management, automatic imposition, job tracking, multi-tasking RIP technology, and other components critical to productivity.

Chromapress consists of four major components: job prep software, server software, a raster image processor (RIP), and the output print engine. Each element is specifically designed to support the system, and the integrated solution is said to be easily assimilated within existing prepress systems. Input is accepted as PostScript files, providing compatibility with a broad range of front-end systems. Job files may be inputted on-line with direct connections through standard network interfaces, or off-line via a variety of storage media as with all digital printers.

Input jobs are passed to a Press Server with workstation. This workstation uses specialized software and off-the-shelf Macintosh hardware to integrate prepress and communications functions, providing complete, centralized system control. Jobs flow from interconnected systems and networks into the server's print queue where they can be tracked, reordered, or canceled.

Both the Print Server operator and/or the original job creator are provided with easy-to-use job production controls to specify variables such as color management, paper type, single- vs. double-sided printing, page order for multi-page documents, and web cutoff length. The Press Server uses this information to process jobs into RIP-ready imposed files. Software automatically notifies the user of any production errors before the high-speed print engine receives the job.

The Chromapress software also supports the industry standard Open Prepress Interface (OPI) file server specification. This allows high-resolution image data to remain in a single location, dramatically reducing network traffic, improving front-end system throughput, and assuring data is readily available to the high-speed RIP.

Jobs from the Print Server are routed to the RIP over a high-speed SCSI interface. This server/RIP interface, combined with centralized file storage, assures that adequate data volumes are delivered to the print unit. This is said to minimize the communications bottleneck normally encountered in short-run printing systems that work with high-resolution color images.

The PostScript-compatible Chromapress RIP was specifically designed by Agfa for this application. It uses a high-speed, multiprocessor, multitasking architecture that supports simultaneous job processing. The RIP also incorporates custom "halftone" screening technology developed for the print engine's variable-density imaging technology. Each pixel can have 64 gray levels by varying the amount of toner deposited on the paper. This increases the available output color range while maintaining image sharpness.

Rapid RGB-to-CMYK and CYMK-to-CYMK color conversions are supported within the RIP as well, eliminating the requirement for preseparated color files. Color management is also integrated. Remote diagnostics are built into the RIP, allowing rapid hardware and software analysis from Agfa technical support as well as "instant" software upgrades.

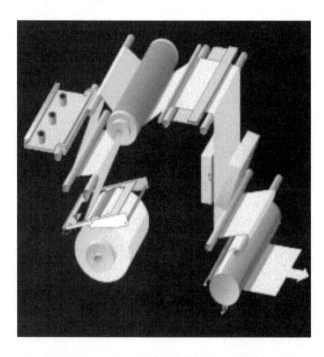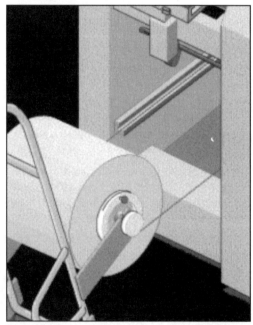

Complete, processed jobs flow from the RIP to the Print Unit over a dedicated, high-speed video interface. Dual job, full-page buffers within the Print Unit use high-speed RAM to maximize page transfer rates from the RIP.

The Print Unit includes a high-speed, four-color, perfecting digital web press. This uses a 600-dpi, variable-dot-density (64 gray levels) electrophotographic output engine. Throughput is 17.5 duplex oversized A3 pages per minute (12.6×17.9 in.) or 35 A4 duplexed pages per minute (8.9×12.6 in.). This is equivalent to 1,050 and 2,100 pages per hour, respectively.

A note about speeds: Most laser printers report their speeds in 8.5×11-in. sheets per minute, single-sided. Since the Chromapress prints both sides at the same time, you could say that it is printing 70 A4 (close to 8.5×11-in.) pages per minute.

The engine uses eight individual color units to simultaneously image both sides of the paper web for perfecting or duplex printing. Each color unit consists of an imaging drum and a tone area. The imaging drum is charged and then exposed with an LED laser. A latent image is formed and attracts fine particles of microtoner. The toner is then directly transferred to the paper. This provides high throughput speeds that are unaffected by the number of colors applied or by duplex printing. The roll-fed paper used within the Chromapress means long print runs — up to two hours with average paper weights, which is about 4,200 duplexed 8.5×11-in. sheets. Agfa emphasizes that there is no sheet feeding mechanism to jam, adjust, or maintain. Further, the system automatically compensates for different paper weight, colors, and finishes

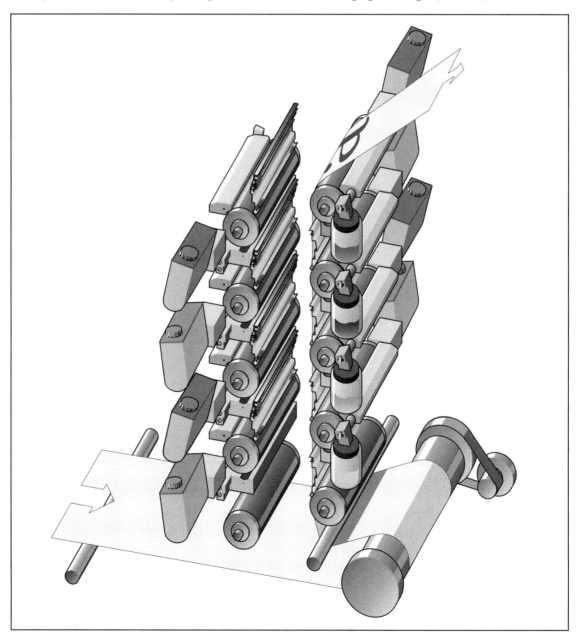

without manual adjustments. Substrates ranging from 40-lb. text to 90-lb. cover weights can be handled. Paper rolls cannot be changed on the fly, but a simple and quick splicing command allows for fast changeover of rolls. Maximum roll width is 12.6 in. (320 mm).

Many paper grades, including coated stock, polyester, and label materials are available. There are currently over 100 qualified substrates. They are available from various distributors including Ressall-Filed and Nationwide.

As the paper leaves the roll and enters the imaging cabinet, it is charged in order to better accept the toner. If the machine is stopped, the paper in the machine loses its charge, and the first ten or so pages are blank.

The print engine images by means of organic photoconductor (OPC) drums using arrays of light-emitting diodes (LEDs) whose average life expectancy exceeds five years. The core LED technology is installed in thousands of Agfa black-and-white laser printers. Chromapress incorporates sophisticated monitoring and feedback mechanisms, insuring consistent light output and, consequently, color across the page, from page to page, and day to day. The stationary LEDs and continuous-paper web are said to ensure excellent intercolor and interpage registration.

LEDs do not emit coherent light as typical lasers do. They are thus harder to focus. Rather than using a scanning laser beam, there is an array of 7,400 LEDs, evenly spaced at 600 dpi, close to the paper. Each LED is responsible for one dot of imaged output.

Agfa developed the specialized microtoners and discrete LED intensities used in the Chromapress to deliver variable printed densities inside each printed pixel, which allows it to achieve a claimed 2,400-dpi and 175-lpi equivalent image quality, which is said to maximize the available color range and image acuity. Dry toners are instantly fused using a non-contact, oil-free process. If the paper web stops, the fuser is immediately pulled out of the way.

The paper web is automatically cut to size with an on-line, automatic sheeter, delivering output to a stacker for use as-is, or for subsequent finishing operations. Sheets can be cut to different lengths.

The Chromapress includes advanced temperature and humidity control systems. Paper is conditioned prior to and following printing, ensuring dimensional stability and process consistency. It is charged prior to LED exposure. The entire print engine is housed in a sealed cabinet about 7.5 ft. high, 6 ft. wide, and 6 ft. deep.

The print unit operates in office surroundings. Toner is replenished using sealed, dust-free, recyclable containers. The nontoxic OPC imaging drums are recyclable as well. All serviceable areas are readily accessible. Chromapress is self-contained, maintaining a clean working environment. Routine maintenance is performed by the operator. About 30 minutes per day of routine maintenance is required, usually cleaning corotron wires.

One of the most important Chromapress advances is the impact on the workflow. Any high-speed, color, demand printing system requires some new skill sets and generally necessitates broader responsibilities for the document producer and/or new skills for machine operators. Chromapress is fully supported by a worldwide Agfa training, maintenance, and support network.

The Chromapress has the ability to vary the printed information of each page on the fly. For each page there are 16 different variable data fields. Text, graphics, and

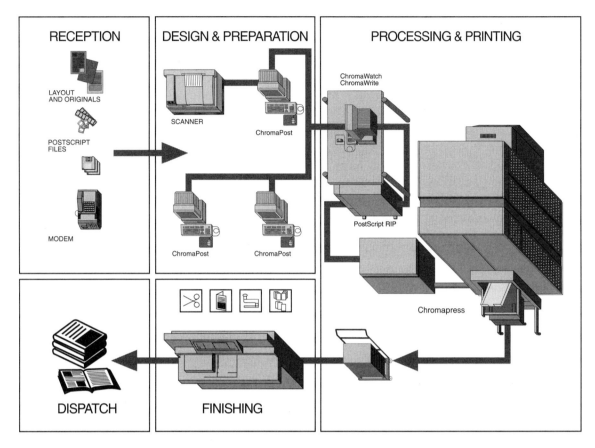

images can be personalized. The black image portion of a complete page can be varied, or approximately 25% of a full-color page can be varied.

The target application is said to be in the 1–5,000 range, but we expect most work to be in the 100–500 area. The basic Chromapress system sells for $360,000 including the software and RIP. Approximately 250 Chromapress systems were installed worldwide as we went to press.

## New RIP

Agfa has announced a new RIP platform that will support the demand for ever-decreasing run lengths and the possibility to personalize every printed page. An architecture of several PowerPC processors will offer at least a quadrupling of the ripping performance.

The AgfaScript PostScript interpreter is fully PostScript Level 2 compatible, and the interpreter offers several quality and productivity enhancements. The RIP is designed for specific Chromapress in-RIP separations with RGB-to-CMYK and CMYK-to-CMYK color conversions, without any speed penalty, facilitating input from different sources without extra preparation time. The raster image processor also allows a default "black overprint" operator setting, thus solving the majority of all trapping issues without using special trapping software. This additional print quality feature does not require any operator skills and does not add an extra step to the workflow process.

## Multipage Support

The Multipage Support feature allows the user to expand the number of pages that can be printed during a production run. The standard configuration supports printing of a maximum of four A4 (or 8.5×11-in.) pages. The optional configurations support 8 or 16 such pages. Depending on the application, the user can choose among three different Multipage Support combinations:

- **Larger formats.** Since Chromapress is a web-fed system and sheets are cut directly on the press, it is possible to print longer sheets. This is of particular interest when printing a 6-page letter-sized or A4 document. This format offers greater flexibility for document design, but traditionally cannot be printed on a small offset press. Printing this type of document in a short run on a large-size press would result in substantial additional costs.
- **Instant job switching.** Successive jobs can be stored in parallel in the engine controller, thus enabling the operator to switch from one job to the next on the fly, without having to stop the engine. In this case the press makeready time is completely eliminated. This feature offers higher productivity for very short runs and therefore reduces printing costs.
- **Instant job proofing.** When successive jobs are stored in the engine controller, the operator can cut in to the next job for a proof, and then go back to the production run. The proof is immediate and does not require the engine to slow down at any time.

## Resident Job Printing

Resident Job Printing allows storage of bitmap information for archiving of previously printed jobs or jobs that have been postponed for printing. Retrieving these jobs from the archive takes only a few seconds. This feature is particularly useful for storing proofs and print jobs that require regular reprints without changes. The maximum configuration allows users to store up to 800 A4 or letter-sized pages.

## Chromapress Software

Agfa has announced that it will sell ChromaPrep, an off-line job preparation software package created specifically for the Chromapress four-color digital printing system. Agfa developed ChromaPrep to meet the needs of the rapidly growing number of Chromapress sites around the world. With ChromaPrep, these sites can maximize their productivity and that of their customers by streamlining the process of handling and printing job files.

ChromaPrep is a complete software suite that allows Chromapress owners to output jobs efficiently and take full advantage of the exceptional speed and quality of the latest Chromapress system. Chromapress customers will use ChromaPrep to preflight files for accurate and fast output on the system, allowing designers, print buyers, and other file creators to ensure that their jobs are printed quickly and accurately.

ChromaPrep includes software components designed to assist at every stage in the digital printing process. It begins with ChromaPost, software that allows users to create a complete job description file specifically for Chromapress. Since ChromaPost supports multiple page-makeup applications, it allows users to work with their current software, without any retraining or other expenses. They simply call on Chroma-

Post to establish specific settings that maximize the quality of output on the Chromapress. With ChromaPost, users can now simulate offset printing when printing with Chromapress — an important capability for outputting a short run on the Chromapress and a longer run on a conventional offset press. These users benefit from the speed and flexibility of digital printing while still matching traditional offset quality.

The Chromapress also includes OnPress imposition software, which automatically handles page imposition, a potentially time-consuming process. Developed with Ultimate Technographics, OnPress automates duplex imposition and supports up to 104 applications. It also allows users to create files for low-resolution imposition proofing, allowing them to check their work before it is output with the Chromapress.

Agfa and Chromapress sites will be distributing the ChromaPrep software.

## New Microtoner and Developer for the Chromapress

Agfa has announced the release of an advanced microtoner and developer that provide the digital color printing system with higher-quality output at a lower cost. Agfa designed the new microtoner and developer specifically for Agfa's Chromapress and Xeikon print engines, including the Xeikon DCP-32D and IBM InfoColor 70.

The new microtoner reduces toner consumption up to 20% per page, while the new developer provides up to three times the life of the previous developer. In addition, these new consumables increase the number of possible applications through increased toner coverage capabilities for each page. The developer's longer life reduces the frequency of periodic maintenance, thereby improving ease of operation and increasing productivity.

Agfa has more than 20 years of experience in both dry and liquid toner technology. After five years of dedicated research and development, Agfa launched dry "color toners" for the Chromapress/Xeikon digital press in 1993. These digital printing consumables, aimed at the graphics market, differentiate themselves from other products on the market by their fine particle size, the evenness of the image gloss after fusing, and the expanded color gamut. Further, the developer offers the longest life of any comparable product used in a production environment.

## OmniGloss

OmniGloss is a high-gloss printing feature that provides a higher-quality, offset-like finish on the Chromapress output. An additional benefit of the new option is that the Chromapress can now handle up to 90-lb. (243-gsm) cover-weight substrates.

OmniGloss provides both controlled gloss and enhanced color saturation through a newly designed two-step fusing system. The new web pull motor, installed as apart of the option, allows use of heavier stocks, a significant advantage when digitally printing covers, business cards, display materials, and more.

## Personalization

Personalizer for Chromapress is an extension of the Personalizer for the Gemini RIP, which is used with Agfa color copiers. Personalizer can merge variable information, black only or four-color, with the contents of a fixed page, called Master Page, during the printing process. The Master Page can be created in any popular page layout program; each side can contain up to 16 variables. During printing, the variable information is ripped on the fly and merged with the Master Page. Personalizer handles the variables independently, which means that they can be changed. As explained in the first section, this enables the user to create extremely targeted documents. Since Personalizer software is fully compatible with both the Chromapress and the Gemini RIP, proofing of personalized documents can be done on color copiers.

## Users Group

North American Chromapress User Association (NACUA) is a joint initiative of Chromapress users and Agfa. As the market for short-run printing is growing and gaining acceptance, and as the distribute and print concept becomes more important, these users groups will grow in importance too. Therefore, the requirement for world-wide print services has come to the foreground. Through the NACUA, a Chromapress user will be able to distribute electronic files to other Chromapress users around the world. The double benefit of this organization is that Agfa's customers can expand their product offerings and will automatically receive more business by providing a service for other Chromapress users.

## Fast Start Marketing and Education Program

To assist users in the development of their digital color printing business, Agfa has established the Chromapress Fast Start Program. As part of the program, Agfa developed several new sales, marketing, and educational tools including:

- **Business Plan ProfitFlow™** — interactive financial software models to help potential users construct a sound business plan and track performance over time.
- **On-Demand Information Strategy Model™** — interactive software designed to calculate the true cost of a printed document, from concept, to creation, to storage, to use, and even disposal. Using this information and actual customer input, a Chromapress user can show key decision-makers, such as communications and marketing managers, chief financial officers, and other corporate heads, how to reduce costs by adopting an on-demand print approach.
- *An Introduction to Digital Color Printing* — a new volume in Agfa's Digital Color Prepress Publications series that explains the essential aspects of digital printing. The guide defines for designer, printer buyers, and potential users what digital color printing is and how it differs from conventional techniques. The guide also provides practical and objective advice and time-savings tips, as well as comparisons of the digital and offset printing output.
- **Chromapress Application Case Studies** — a guide that highlights strategies being implemented by successful Chromapress users that may be of interest to users seeking out new markets and customers. Each guide includes a description of numerous real-world applications and printed samples of each project.
- *Hot Off the Press* **Public Relations Kit** — a focused customer public relations guide book developed exclusively for Chromapress users to help them promote their Chromapress services. The guide provides customers with recommendations, ideas, and detailed instructions on how to launch and manage their own public relations activities.
- **Chromapress Digital Resources** — a digital library containing various images, graphics, and marketing material relating to Chromapress digital color printing.

# Chapter 29

# IBM Printing Systems Company

The personal computer printer market, like other computer product markets, has been influenced by the breakneck, nonstop pace of the computer chip industry. As a result, companies competing in that market, such as the IBM Printer Division, have gone through radical changes.

Arguably the personal computer revolution began when the Apple II was unveiled at the first West Coast Computer Faire in San Francisco in 1977. But it wasn't until the IBM PC was introduced 1981, with an open architecture along with Lotus 1-2-3 (announced one year later), that PCs became legitimate in the computer world.

Although appearing beneficial for IBM at first, the shift to open smaller personal computer systems, along with greater competition in all of IBM's segments, caused wrenching changes. After posting profits of $6.6 billion in 1984, IBM began a slow slide. The company began reducing its work force in 1986, and by 1992 it cut worldwide employment by 100,000 through attrition and early retirement inducements.

In the early 1990s the IBM printer products were split into two units: Lexmark and Pennant. Lexmark sold smaller units, individual printers. Pennant focused on the highspeed workgroup printers. As the slump continued, many divisions were sold: the copier division to Kodak, 1988; Rolm telecommunications to Siemens, 1988; and the typewriter, keyboard, personal printer, and supplies business of Lexmark, to Clayton & Dubilier.

The Pennant division, often associated with the AFP data stream (printer language), evolved into the IBM Printer Division in 1993 and today is called the IBM Printing Systems Company.

While these hardware changes occurred, software and printer language changes were occurring. Ironically, both the Adobe PostScript language and IBM's AFP language were introduced around the same time in 1984. PostScript focused on single-page output and the graphic arts community. AFP was designed for multiple-page output and the corporate and in-plant markets. These two output languages progressed on different paths until the 1990s.

For the first few years, IBM and Adobe moved in similar directions as both became interested in portable document formats (PDF) or "viewer" technologies. As Adobe refined its PDF format, it recognized the importance of multiple-page format, error detection, and data recovery. During the same time period, IBM recognized the dominance of the PostScript and HPCL languages as printer sales shifted from a centralized workgroup to a decentralized individual environment with one in everyone's office.

As a result IBM has embraced the PostScript language, as evidenced by IBM's new alliance with Adobe. At the 1995 Seybold show in Boston, IBM announced a collaboration with Adobe to bring AFP functionality to the desktop. "To IBM," said Dr. James T. Vanderslice, former general manager of the IBM Printer Division, "print-on-demand represents a whole new business beyond our traditional printer business."

These recent changes do not mean that IBM is abandoning AFP. A better explanation is that IBM is willing to build bridges "across to" and "across from" the PostScript language. In some ways it is analogous in strategy to the way Xerox incorporated PostScript and HPGL into the DocuTech. Xerox's internal printer architecture remained Interpress, but Xerox bridged the gap between Interpress and PostScript and HPGL.

## AFP and Corporate Support

The core of many of the IBM Printing Systems legacy products is the Advanced Function Printing (AFP) family of software and hardware, which support the distribution and presentation of information — in multiple forms, across multiple platforms for displaying, printing, and storing on multiple types of devices.

For in-house or corporate users who are already using AFP, IBM continues to support them with a wide variety of network operating systems, network protocols, printers, and application data streams. AFP applications can be distributed across a variety of environments including System/390, AS/400, AIX, and PS/2 environments as well as local- and wide-area networks.

AFP supports two of the common network operating systems, Novell Netware and IBM LAN Server. The Ethernet and TokenRing network protocols are both supported as well as the following printer data streams: Intelligent Printer Datastream (IPDS), Page Printer Datastream (PPDS), and Hewlett-Packard Printer Control Language (HP PCL 4 and 5).

AFP's support of industry standards extends beyond networks and printer data streams to include the most popular application platforms and data streams. The platforms supported include MVS, VM, VSE, OS/2, DOS, DOS/Windows, OS/400, and AIX. Any platform that generates accepted data streams is compatible. These include AFP Datastream, ASCII, Postscript Level 1, MetaFiles, TeX, and Ditroff.

## Print-on-Demand Solution

IBM is focusing on complete print-on-demand (POD) solutions. These solutions will include computer hardware and software as well as the output devices. At the time we go to press, the first solution is for the 3900 family, but additional printers will be supported in the future.

Here's how it works. Users create POD files with typical DTP applications (i.e., QuarkXPress, PageMaker) and create job tickets using the job ticket software known as the POD Operations software. The job ticket contains information about the job such as the printing parameters (one-sided or two-sided printing), paper stock, and binding specifications, as well as the files and resources required to print the job.

Specific job parameters can be set up and stored in files or input as needed. Files can be sent from Mac and Wintel (Windows/Intel) computers.

After the job ticket is created, users submit the files to the POD Scheduler. The POD Scheduler is a user-friendly graphical interface that allows the user to change the priority in the printing queue. The jobs are rasterized with an Adobe PostScript RIP, and then printed to a high-speed printer (i.e., 3900).

The current system uses the 3900 printer (described in the next section) and a RS/6000 Multiple Printer Controller. This RS/6000 can operate two 3900 printers. The POD Scheduler, POD Library database, and Adobe RIP operate on the RS/6000.

## 3900 Family

The 3900 printer family can print from 6 to 11 million impressions per month. The products in this category include the 3900 Advanced Function Printer, 3900-Wide Advanced Function Printer, 3900 Advanced Function Duplex Printing System, and the 3900 High Resolution Printing System.

On all units, the paper is fed from a continuous roll (web) and the imaging technology is laser electrophotography. All 3900 engines are driven by PowerPC technology running AIX.

These printers can be purchased as either a simplex or duplexed version. The duplex printer comes with two units operating at 229 ppm, each resulting in 708 two-sided impressions per minute. The resulting pages can be a maximum 18 in. long and 17 in. wide.

Print quality is high due to a newly developed system and the IBM exclusive Print Quality Enhancement. Together these enhancements provide darker black fills, smoother edges, and boldness control.

The 3900 Advanced Function Duplex Printing System can print on paper that is 17 in. wide, allowing users to print two-up, 8½×11-in. pages. For this unit, the server is a PowerPC-based Advanced Function Common Control Unit (AFCCU) that can operate at 708 ppm.

A wide variety of resolutions are available. Most printers in the 3900 series are 240 dpi, a few offer 300 dpi, and one, the 3900 Advanced Function Duplex Printing System, can be configured for 600-dpi output.

First shown in Phoenix at the Xplor show in November 1994, the 3900 Advanced Function Duplex Printing System continues to evolve. At the On Demand show in New York City in April 1996, it was offered in print speeds of 480 ipm for the simplex model and up to 708 ipm for the duplex model, both in two-up format.

A variety of finishing solutions were also demonstrated at the show. Roll Systems, Inc., an IBM Business Partner, featured its paper system for the IBM 3900 duplex printer, which introduces significant savings in both paper costs and waste handling. Finishing solutions from Stralfors Lasermax International Inc., C.P. Bourg, Standard's Finishing Systems Division, and Gunther International were also be demonstrated.

## IBM InfoColor 70

The October 24, 1995 announcement that IBM would resell the Xeikon DCP-1 surprised many people within the graphic arts community. The surprise was the result of the fact that the Xeikon engine was already being distributed in North America by both AM International and Agfa as the Chromapress. However, as we discuss later, IBM brings some unique technology, marketing, and support advantages to the table.

Originally called the IBM 3170, the InfoColor 70 is based on the Xeikon DCP-1 four-color print engine. It is a 600×600-dpi device using a light-emitting diode (LED) print head and electrophotographic technology. A broad color gamut can be produced through the use of a small-particle-size, translucent, dry toner. Color matching is targeted for SWOP or Euroscale.

The internal environment of the InfoColor 70 is continuously adjusted to optimize print quality. For example, humidity and temperature are monitored and controlled. Paper is air-conditioned before it enters the imaging stations. The InfoColor 70 also varies the electrostatic charge to produce consistent toner density in full color. Color matching becomes simple and consistent from piece to piece, and run to run.

The RIP technology uses proprietary screening technology to produce halftone images at a screen frequency of 170 lines per inch (lpi), with a classic clear-centered rosette pattern. It uses dedicated Application Specific Integrated Circuits to produce clear, sharp images that approach offset press quality.

Of course, the underlying engine, the DCP-1, accepts PostScript data streams from an Ethernet interface, and IBM has announced that it will add the ability to print with its AFP language.

## Unique IBM Advantages

Considering that the Xeikon-based devices are available from a number of OEMs, a natural question is what differentiates the products delivered by these OEMs. Each manufacturer offers unique advantages. Let's examine the IBM advantages.

With years of experience with electrophotographic printing devices, IBM has engineering experience as well as an established team for service and support. IBM's market focus on Fortune 1000 companies and in-plant reprographic departments provides IBM with a different market and distribution strategy than the competing OEMs, which focus on the graphic arts.

With 2,000 customer engineers dedicated to printers in the U.S. and 4,000 worldwide, IBM has a strong network for support. According to IBM, most of the engineers have between five and sixteen years of experience in electrophotographic technology.

The maintenance program offered by IBM is unique. The cost of the consumables is paid through a "click" charge. IBM claims that this is superior to the alternative, which is allowing the user to buy the consumable because the actual life of the consumable often is less than the expected life. Therefore the user pays more for the consumables than they expect. By charging for the consumables within the maintenance agreement, IBM is guaranteeing a stable price even if the consumables expire prematurely.

Under the standard maintenance contract, IBM offers support 24 hours a day, seven days a week, via a toll-free phone number. According to a competitive service analysis, the standard support agreement of other OEMs is for one shift, five days a week, and some do not offer 24-hour support at all.

In addition to its support of larger paper rolls, the IBM Printing Systems Company is making its extensive paper testing lab available to assist customers with qualifying media for their applications. Customers can send paper to IBM for testing, and IBM will develop script files to help automate the paper conditioning controls on the InfoColor 70.

As with all IBM printers, customers will be supported by the industry's largest dedicated team of printer specialists, who will provide high-quality on-site assistance. The IBM Printing Systems Co. boasts 4,000 specialists worldwide, including 1,800 in the U.S., providing service 24 hours a day, seven days a week.

## Unique InfoColor 70 Enhancements

IBM performed an engineering audit of the Xeikon DCP-1 engine before signing on as a OEM. The audit resulted in hundreds of engineering suggestions. Many of these improvements have been implemented into all of the Xeikon engines and some into only Xeikon engines destined for IBM. One example of this change made specifically for IBM is a heater that IBM added to improve the print quality for a specific range of humidity and temperatures. Other examples are:

- Color matching tables and SWOP and Euroscale standards
- Automated density control
- Clarified operator error messages
- Unique variable data implementation
- POD software

The IBM method of managing color is unique. One approach to matching colors is to print a swatch on an ongoing basis and then enter the values into the RIP to keep color constant . IBM found that this is time-consuming and the color always changes depending on what paper is being used. Instead they opted to write their own lookup tables (LUTs).

Also IBM analyzed the color output and standardized the color to match the SWOP (specification for web offset press) and Euroscale standards. All color is then matched at the desktop level using International Color Consortium (ICC) profiles. This should reduce operator intervention and monitoring time.

The Automated Density Control automates difficult, time-consuming operator adjustments that are the key to excellent and consistent print quality. The InfoColor 70 densitometer, which is plugged directly into the IBM RS/6000 PowerPC, analyzes data and changes the source code accordingly. It automatically changes perimeters instead of the traditional approach where the operator reads the swatches, writes down the data, evaluates what their perception is, and makes the change on the press.

The InfoColor 70 gives operators error messages in easy-to-understand "plain English," rather than numbers and codes, and instructions to direct the operator on how to proceed. This enhances troubleshooting and increases productivity.

As far as we know, IBM is the only OEM vendor that has a dedicated paper testing lab where customers can send paper for testing.

The IBM variable data creation tools allow document owners to create and implement variable data into their documents (using QuarkXPress and PageMaker). The tools are easy to use and have some unique features. For example, IBM leaves the variable data separate from the master data until the job is ripped. Also the Merge-Doc variable data production tool preps and checks variable data jobs for the RIP.

## InfoColor 70 and Variable Data

By integrating IBM Printing Systems Advanced Function Presentation (AFP) architecture, the web-fed InfoColor 70 lays the foundation for customers to personalize each document by incorporating variable data from existing databases.

At the On Demand show on April 23, 1996 in New York City, IBM announced the availability of personalization software called Variable Data System (VDS). VDS allows the InfoColor 70 to print variable information such as text, graphics, and images in both black and/or full color. VDS manages high-speed PostScript printing from Macintosh and Windows-based clients.

The software, which runs on RISC System/6000 and servers under the AIX operating system, controls local and remote print submission, job ticketing, and job management. Customers may also choose to install an optional RS/6000-based library server for data on-demand.

According to IBM, printing with VDS does not slow down the IBM InfoColor 70. Customers can change up to 50% of the data in full color and 140% of the black on both side of every duplex page while the printer is running at 35 pages per minute, 70 ipm.

Another new feature is the Electronic Collator that electronically arranges the pages in finished sets before printing, eliminating off-line collation. In contrast traditional collators are mechanical devices.

The collator is capable of storing up to 32 GB of data, equivalent to about 500 pages of a typical magazine. This capacity enables the InfoColor 70 to print collated documents with full variable information on A4-size pages at 70 ipm. With text-only documents, the collator can store from 5,000 to 10,000 pages. The Electronic Collator feature also extends variable data to the entire page — 100% of both sides can change from page to page.

## Xplor Award

Xplor, the Electronic Document Systems Association, is the worldwide association of nearly 2,000 organizations whose people are responsible for the management of strategic information using the technology of the $75-billion global electronic document systems industry. On November 9, 1995, IBM and Xeikon jointly received the "Innovator of the Year" award for the InfoColor 70 (IBM 3170) from Xplor International.

"It is exciting to have a short-run, full-color digital printer in our product line. This printer is the flagship product of our print-on-demand strategy, and it is the catalyst that is unifying the formerly disparate worlds of production printing, distributed printing, and print-on-demand," said David R. Carlucci, general manager of IBM Printing Systems Company.

"We chose to give this prestigious award to IBM for its 3170 printer because it exemplifies the innovation that arises when leading industry vendors partner to produce ground-breaking new products," said Ray Simonson, president of Xplor International. "Technology such as the IBM 3170 gives Xplor members new opportunities to exploit the color print-on-demand market."

## Summary

Although coming as a surprise to may people in the graphic arts community, the distribution and sale of the Xeikon-based IBM InfoColor 70 makes sense. IBM has a strong distribution channel into the corporate market whereas AM International and Agfa focus more on traditional graphic arts companies. Consistent with their strong distribution channel is their service and support abilities.

With years of engineering and manufacturing experience creating and servicing electrophotographic printers, IBM has the resources to build unique enhancements into its printing engines. Also, with its AFP output language, IBM has as much experience as any manufacturer in implementing variable printing.

# Chapter 30

# Indigo E-Print 1000

Indigo was founded in 1977 by Benzion "Benny" Landa, the company's chairman and CEO. In 1992 Indigo went public. According to *Israel Business Today* (July 23, 1993), the event that triggered the stock offering was when international financier George Soros (Soros Venture Capital) bought $50 million in stock (15%). After that event, Indigo Graphic Systems was founded in 1992 with funding from First Boston Corporation, Toppan Printing of Japan, as well as other financial backers. The balance is held by Landa and his family.

In February 1996, Wayland Hicks, former executive vice president of operations for Xerox, and vice chairman and CEO of Nextel Communications Corporation, was named president and CEO of Indigo N.V.

Landa creates as much press as his Indigo E-Print 1000+ Digital Offset Color™ press. He holds over 100 patents, including those for ElectroInk®, Indigo's unique ink technology. "Indigo," says Landa, "has invested 16 years pursuing a single goal: merging the quality, economy, and performance of printing ink with the power of electronic printing." Landa claims that the Indigo's Digital Offset Color is "the only process that combines the unique qualities of liquid ink with the durability of the offset process."

Until the E-Print 1000 digital color press was introduced, Indigo was better known in the office copier market than in the professional graphic arts market. It has more than 200 U.S. patents and hundreds more worldwide in copying technology. According to Indigo, it is hard to find a copier that doesn't include at least one Indigo patent.

Ironically, just as a number of senior staff from Xeikon came from Agfa, a number of senior staff from Indigo came from Scitex. For example, Shlomo Waldman, the general manager of operations, was formerly the director of manufacturing plants at Scitex, and Steve McClean, the vice president of sales and marketing, was formerly the head of sales with Scitex.

With corporate headquarters in the Netherlands, worldwide marketing operations centered in the United States, and research, development, and manufacturing operations based in Nes Ziona, Israel, Indigo is truly an international company.

## Finances

As we are going to press with this second edition, we learned that Indigo reported total revenues of $28.3 million and product revenues of $27.3 million for the first quarter of

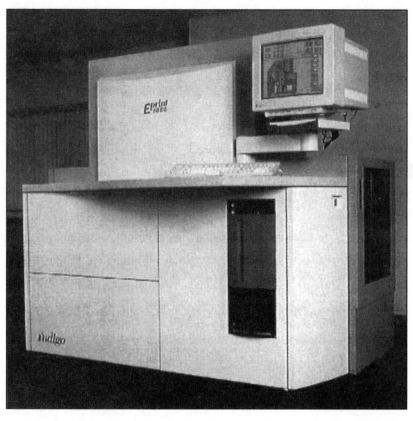

The Indigo E-Print 1000.

1996 compared to total revenues of $42.6 million and product revenues of $40.7 million for the first quarter of 1995. The company reported a loss per share of $0.40 for the first quarter of 1996 and a loss per share of $0.01 for the first quarter of 1995.

Gross margin as a percentage of product sales decreased significantly to 16% for the three months ended March 31, 1996, compared to 46% for the three months ended March 31, 1995.

Selling, general, and administrative expenses increased 21% to $21.2 million for the first quarter of 1996 from $17.6 million in the first quarter of 1995, primarily due to the expansion of the company's direct sales and marketing channels in Europe and North America. Selling, general, and administrative expenses declined 12% to $21.2 million for the first quarter of 1996 compared to $24.1 million in the fourth quarter of 1995 due to a reduction in marketing expenses associated with trade show activities.

According to Indigo, the company's total revenues for the first nine months of 1996 is $80.6 million, compared with $128.6 million for the same prior-year period. The company said that despite the reduction in total revenues, the operating losses were down slightly, compared with the second quarter of 1996. Revenues from imaging products and services continue to increase. The average monthly impressions per installed Indigo press increased during the first nine months of 1996 by approximately 50%, both in Europe and the U.S. On a per-share basis for the nine-month period, the company reported losses of $1.08 in 1996 versus $0.42 in 1995.

In May 1996 the company completed a private placement of $108 million of equity to a limited number of institutional and private investors including Chancellor Cap-

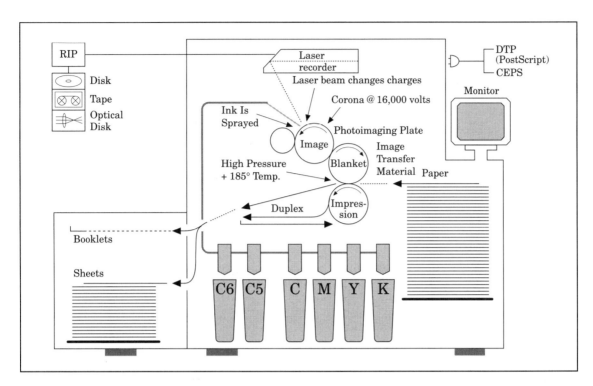

The basic principles of operation of the Indigo E-Print 1000.

ital Management, The Renaissance Funds, Janus Capital Corporation, as well as the company's two principal shareholders, an affiliate of the Soros Group and a company owned by a trust of which Benny Landa and his family are beneficiaries.

## Basic Technology

Indigo began developing its true, fully digital color press a decade ago and is reported to have invested more than 2,000 person-years and more than $150 million to bring it to market. The technology utilized in the Indigo was under final development for at least three years before it was first introduced on June 22, 1993.

The E-Print is a sheetfed digital offset press that prints four to six colors on most of the popular paper stocks at 800 dpi, 11×17 in. (279×432 mm) in size (A3), at 4,000 sheets per hour, at one color on one side, or 67 per minute. Thus, for black-and-white work, it is just about the same speed as a Xerox DocuTech (135 8.5×11-in. pages per minute).

It is 6 ft. 3 in. (1.9 m) long and 4 ft. (1.2 m) deep, taking up only 20 sq.ft. Compared to a traditional press, the operation is nearly soundless. Although originally marketed as a press that could accept paper stock from almost any source, the erasability issue required a substrate surface pretreatment, which tended to limit the paper sources. Today, there is a wide array of optimized paper, film, and plastic substrates available from major manufacturers for use on Indigo digital presses.

It is not clear exactly how it was discovered, but only a short while after the Indigo was introduced in the United States someone, perhaps Mike Bruno, world famous printing expert, took out an eraser and tested the erasability of the ink. He found as did many other people that you could erase the Indigo ink. As we have discussed in the paper chapter and will discuss in more detail later, Indigo's ElectroInk does not adhere to paper like conventional inks. Figuratively speaking, the ink is "laminated" into an ink or a plastic film, and then it is peeled off the blanket and applied to the substrate paper with the help of the transfer oil.

Paper substrates for Indigo presses can be coated or uncoated, varying in thickness from very light sheets to card stocks. The E-Print 1000+ can print one color on 4,000 A3 (11×17-in.) pages per hour, or 67 per minute. Up to 3,000 sheets can be loaded in the feeder, though single sheets are handled separately.

In some ways, it uses the same principles as traditional offset printing, with plates, blankets, and impression cylinders and liquid ink. The image on the plate cylinder is transferred to the blanket surface and then is "offset" to the paper held on the impression cylinder. However, in other ways, the Indigo digital press works like a copy machine by charging a cylinder and using a laser beam to create an electro-photographic image.

As a sheet is finished, it moves to the Booklet Maker where booklets are automatically gathered and folded, stapled, and stacked. From the job setup, the Booklet Maker knows how many sheets to grab to make the booklet, whether to staple, etc. Thus, the E-Print's final product is the finished piece itself.

## Paper (Substrate) Update

As discussed earlier in the paper chapter, there have been some paper issues when using the Indigo press. However, the Sapphire pretreatment is working well, and the number of available paper manufacturers has increased significantly.

These pretreated papers are now marketed by leading manufacturers, including Champion International, S.D. Warren, MACtac, 3M, PPG, Fasson, Mobil, Folex, Wyndstone, Hanita, and Bell Industries. Indigo is continuing to work closely with these and other manufacturers to continue the expansion of Indigo-optimized substrates available in the marketplace. Currently, 37 Indigo-optimized substrates are in distribution for the E-Print 1000+, and 39 optimized substrates are available for the Omnius™ press.

The industry's leading substrate manufacturers understand the tremendous potential of the short-run, color printing market being driven by the growing use of Indigo digital presses. "With our comprehensive portfolio of Indigo-optimized substrates readily available, our customers are creating new and imaginative applications for on-demand digital color printing. The substrate partners we have chosen to help us develop the on-demand, color digital printing market are all recognized industry leaders. Through their creativity and innovation, we are accelerating the acceptance of digital printing. It is a win-win situation for the substrate producers and Indigo."

The wide and increasing availability of Indigo-optimized substrates is the direct result of Indigo's ongoing strategy to expand the short-run on-demand digital color market by providing the most complete palette of substrate options for all types of end-user applications.

The selection of Indigo-optimized substrates in the distribution pipeline today encompasses a wide range of coated and uncoated papers, pressure-sensitive label stocks, films, and even heat-transfer backings. Indigo owners can offer their customers a wide choice of substrates for their applications. Short-run applications for coated and uncoated paper substrates include high-quality, full-color sales brochures, sell sheets, direct mail inserts, newsletters, personalized target marketing materials, and customized proposals.

Indigo-optimized pressure-sensitive substrates are being used for printing full-color variable-image labels for the growing short-run label market. In packaging applications, Indigo-optimized films are used to produce colorful flexible packaging for consumer products.

Plastic film substrates are used to produce color overheads for textbook publishers. Heavier, durable plastic substrates are being utilized for telephone and courtesy cards, as well as identification products like nameplates, barcode plates, outdoor signage, and displays. In addition to leading substrate manufacturers, Indigo has developmental alliances with several other key companies in the electronic imaging, communications, and printing industries, including Adobe Systems and EFI.

## Computer

The basic E-Print digital press comes with a Sun-based workstation, a hardware interface and a print controller buffer containing between 32 and 640 MB of RAM. Indigo uses its own image compression technology, which achieves a 5:1 compression ratio. This compression causes a loss in the image fidelity but allows the E-Print to store more pages. Indigo's marketing director, David Leshem, calculates that 100 pages of text, averaging a couple of typical magazine color photos on each page, will fit within the buffer.

The Sun graphical user interface gives the operator three choices: load, process, and print. There are several advantages of the Sun-based systems. The Sun platform using the UNIX-based operating system offers true multitasking. Therefore all three tasks on different jobs can be performed simultaneously, minimizing idle time for the press. Another advantage is that faster Sun computers can be added to perform the jobs faster.

The load function pulls in desktop pages or color data into the Sun station and converts it to Indigo's internal format. The process function automatically rasterizes individual elements and pages, and electronically collates them for rapid imaging.

To image 8,000 A4 pages per hour, the Indigo's print controller must produce 200 Mb/sec. This is beyond the capability of today's networks. Therefore the chore is divided between the Sun workstation, which performs the loading and rasterizing, and a dedicated controller with a 640-Mb raster image. The two are connected by an S-bus card and a vme/vsd-bus interface, which can achieve a maximum data transfer rate of 200 Mb/sec.

At the input station, the operator manipulates each individual print job by adjusting color brightness and saturation, color correction curves, and the document composition (moving individual elements of a page, adjusting imposition, or finishing).

The storage options include additional hard disks, optical discs, and Exabyte tapes. In addition, the E-Print can also be on-line to a local-area network. Future options will probably include Photo CD.

If the print job exceeds the RAM buffer, the E-Print 1000+ stops and waits until the next page is ripped and the memory is flushed. On the other hand, if the print job does not exceed the RAM buffer, additional jobs can be held in the print queue.

## RIPs

The E-Print 1000+ prints PostScript files from a Mac or PC as well as Scitex's internal LW and CT formats via Handshake GPIB or Exabyte tape. The RIP can be in-line or off-line. The off-line RIP, called the E-RIP, allows greater overall throughput.

The E-RIP is an Adobe CPSI Level 2 interpreter, plus the Indigo software, running on a Sun workstation with a gigabyte (GB) or more of disk storage and Ethernet and TCP/IP networking. In addition to rasterizing, the off-line workstation has software for inputting job layout and parameter definitions.

Rational screening technology is employed with line screen frequencies of 133–150 lines/in. (5.2–5.9 lines/mm). According to the rule-of-thumb calculations of dpi input required to lpi output for screening, there isn't enough resolution to support these rulings. However the Indigo printing technology is closer to inkjet technology than to offset technology.

Other formats, platforms, and protocols have been announced but not delivered. Crosfield and Linotype-Hell's internal format will be supported. Other input options include tape, optical disc, and local-area network. Networks supported will be NFS, TCP/IP, IT8.4, Novell, Lantastic, and DECnet. Formats supported will be TIFF, DDES, Handshake, Targa, and EPS/DCS. Another advantage of the Sun workstation is built-in Ethernet connectivity.

## Reusable Dynamic Plates

The printing function starts imaging the reusable digital electrophotographic offset plate. While the plate is imaged electrophotographically, as is commonly done with copying machines, the E-Print 1000+ plate is reusable. Indigo, which calls the electrophotographic plates "dynamic," says they last tens of thousands of impressions each.

The plate cylinder is electrostatically charged and exposed with the laser as it turns at a speed of 4,000 impressions per hour. Next the cylinder rotates and is exposed to ink spray of one color (cyan, magenta, yellow, or black) from the ink nozzles called the Ink Color Switch. As the plate cylinder continues to revolve, the ink is transferred to the blanket and then all of the ink is transferred to the paper.

## ElectroInk

Besides being able to expose and image on the fly, the press is designed such that a completely different image can be created with each revolution. This is enabled by the reusable plates and by the inks. The 100% ink transfer and reusable plates allows each image to be different. In color printing, it could be the next color separation, while in database printing it could be the personalization of text and images on each page. In four-color printing, it is the Ink Color Switch that changes the ink color.

The ability to transfer 100% of the ink from blanket to substrate is unusual in traditional offset printing. This is made possible with the E-Print because of Indigo's patented liquid ElectroInk, which uses pigments similar to those in regular offset inks but with two dramatic differences. First, it acts electrostatically, meaning it can be charged, and second, it dries very quickly.

Contained in the inks is a dispersion of pigmented polymer particles ranging in size from 1 to 2 microns. In contrast, the dry toner particles used in copy machines have an average size of 8–15 microns. When transferred to the blanket and heated, these polymers turn into a tacky polymeric "film."

When the ink film polymer comes in contact with the substrate, it hardens instantly and peels away from the blanket. There are two interesting contrasts with traditional printing. First, with the E-Print 1000+, the ink does not bind with the substrate paper as in traditional printing but laminates or coats the substrate. Second, with the E-Print 1000+ press, 100% of the ink is removed with each revolution, while in the conventional offset printing process, half of the ink is transferred to the paper and the balance stays on the blanket, to be re-inked on the next revolution.

**Advantages.** There are several advantages of the Indigo's ElectroInk technology. It allows for individualization or personalization of each page. The small particle size results in a printed product that feels more like offset than the raised image from copying technologies. And lastly, Indigo claims that this process results in sharper images.

According to Indigo, the printing produced by the E-Print 1000+ press has better edge definition, or acutance. The acutance is higher because there is no wicking or bleeding of the ink as it hits a paper substrate, which occurs in a conventional wet ink transfer. The dots show no feathered edges, and there is no dot gain because the ink doesn't flow onto the paper; it bonds to it.

Indigo claims that show-through to the other side of the paper is minimal for the same reason. Yet, at the same time, the ink film is so thin that it doesn't appear plastic, as with thermal proofers. Indigo explains that the ink film is thin enough to replicate the texture of the paper fibers because the ink particles are so small.

**Disadvantages.** The main disadvantage of the ElectroInk technology is the erasability issue. In conventional printing, the ink dries by absorption, evaporation, or heat, while in the xerographic printing process, toners dry by heat or pressure fusing them into the paper. The drying process for ElectroInk is different. Figuratively speaking, the ink is "laminated" into a ink-plastic film, and then both ink and film are peeled off the blanket and applied to the paper with the help of the transfer oil. This occurs for each of the four process colors.

In conventional printing, the ink binds with the paper. With the ElectroInk, the ink dries to a film for transfer from the blanket, before it reaches the paper. The ElectroInk does not penetrate into the paper, and thus the ink has a lower degree of adherence.

Although originally dismissed by Indigo, which said that none of its customers or prospects had withdrawn from contracts on the basis of erasability, it has recognized the problem and developed a "work-around" solution that increases the adherence of the ink. The solution, called "Sapphire," is a pre-treatment of the paper with a chemical. According to company officials, this process is not unlike to running the paper through a press with the water dampening on. This pretreatment process may work by allowing the paper fibers to rise off the paper so the ElectroInk in the ink film stage adheres to the fibers.

Another issue in the operation of the Indigo is ink supplies. Currently only Indigo makes the ElectroInk in Japan (with Toyo Ink). Is ElectroInk made in North America and Europe? The issue, of course, is that competition drives prices down.

Although not a disadvantage, while discussing the inks, it should be noted that they can approximate SWOP and other process color standards. The inks are made in a light mineral oil base and come in sealed canisters that many users refer to as "spray cans."

## Print Speed

It is difficult to compare E-Print 1000+ speeds to conventional press speeds. Conventional presses print one, two, three or four colors at the same speed (on a multicolor press). In contrast, the E-Print 1000+ prints one color pages faster than two, two colors faster than three, etc.

This ability is due to the very tight press registration. The paper is held in place on the impression cylinder throughout the imaging process for each side of the paper.

Once all of the colors have been printed on the first side of the page, the sheet is transferred from the cylinder into the duplex buffer. Next the trailing edge of the paper is picked up and pulled back to the impression cylinder such that the other side is presented to and retained on the impression cylinder. At this point the colors can be applied to the reverse side of the sheet.

The current maximum E-Print 1000+ operating speed is 4,000 A3 impressions per hour, single sided and single color, or 8,000 A4 pages/ hour, which is equivalent to 133 A4 pages per minute. Interestingly this is about the same speed as a Xerox DocuTech, which prints at 135 A4 pages per minute. In this specific example, both are printing in black and white, but the DocuTech resolution is 600 dpi, while the resolution of the E-Print 1000+ is 800 dpi.

To calculate four-color pages we need to divide the 4,000-impression rate for 11×17-in. (A3) pages by four (for four colors), which results in 1,000 impressions per hour. For duplex four-color printing the speed is cut in half, to 500 impressions per hour. The forthcoming Indigo Mobius digital press will be faster.

## Bindery Capabilities

Using the electronic collation and the Booklet Maker option enables completion of booklets without manual intervention. The imposition and the job setup specifications are input on the Sun workstation. The collation is handled in the controller during the print process.

At the delivery end of the press, the finished pieces come off either as sheets or as folded and stitched books containing a maximum 100 pages each. This is quite different from conventional sheetfed or web presses, which print the same image over and over in succession. In the conventional sheetfed press workflow, collation is performed in the bindery process.

One of the unique issues for the Indigo press is data storage. Since all the pages for a single piece are printed together, in succession, all the information must be stored. Maintaining and utilizing all the information over and over requires a significant amount of memory and fast processing.

A 640-MB RAM buffer stores this information. If one particular job does not take up the entire 640-MB RAM buffer, additional jobs can be queued while printing proceeds on the first job. Using the Booklet Maker, however, requires that the entire job be stored in memory throughout the print run.

An additional advantage of the RAM buffer is that it is useful in the process of personalization or customization. Page elements that repeat from page to page are stored in the RAM buffer in a rasterized form. In this workflow, only the variable data has to be identified and loaded from the database, in a rasterized form.

## Comparatively Speaking

Indigo must compare itself to conventional wet ink printing as well as powdered toner printing. On the wet ink side, Indigo claims several advantages: Show-through is minimized since there is no water. Dot gain is virtually nonexistent. The ElectroInk is very thin and bonds to the paper. Each color is translucent. Paper shrinkage, curl, or other artifacts of wet ink printing are not a factor.

Against powdered toner technology, Indigo claims a quality advantage: the 800 dpi provides excellent typographic quality.

To duplex-print, the sheet is released by the impression cylinder after it is printed into a duplex buffer. The trailing edge of the paper is clamped back onto the impression cylinder and the second side is printed. Each side can be printed with as many or as few colors as desired. The printed sheet is then ejected. An optional Booklet Maker retains the printed sheets (up to 100 pages) and then releases them to an on-line folder/stapler. There is no trimming.

Once again, speed is difficult to completely measure. Once rasterized, pages are sent to the Sun workstation in the printer, and about 100 pages with a some photos on each can be buffered. At that point the printer operates at its full speed, printing a combination of full-color and black-and-white pages as well as some with spot color. See our speed comparison chart later in the report.

The first user of the E-Print is Pioneer Plates, Ltd in Holon, Israel. The company is charging the equivalent of $1.00 per 11×17-in. (A3), which is very competitive with a Canon CLC and EFI Fiery combination.

For variable data, the E-Print could expand on its ability to produce a completely different page with every rotation of the cylinders. The software and supporting network technology must be introduced, and we expect it is a primary development effort.

The E-Print accepts PostScript Level 2 data and Scitex Handshake data, and will soon accept PostScript Level 2 data as well as other formats. An agreement was announced with Adobe Systems to support Adobe's new SUPRA architecture.

The unbundled list price of an Indigo E-Print 1000+ digital color press is $379,000, but a fully-configured system with the six inks and the Booklet Maker would approach $450,000. Indigo claims to have sold out its entire first year of production of 30 units per month, or 360 systems. Deluxe Check Printing is said to have ordered over 100 units. Moore Business Forms, both U.S. and Toppan Moore in Japan, is the next largest customer. The majority of customers have been larger commercial printers. Most of the early adopters ordered more than one machine.

Toyo Ink of Japan will also be manufacturing the ink. Indigo's U.S. operation is based in Massachusetts.

## Drupa 1995 Announcements

The E-Print 1000+ enhancements provide significant improvements in reliability, quality, and performance. The print size has been increased to 11.7×17.2 in., giving full A3 page production including bleed. Improvements in the color algorithms and the page handling system give substantial improvements in image quality, color control, and overall image registration.

Advances in image quality for the E-Print 1000+ are provided by High Definition Imaging, an option that provides an effective line screen of 200 lines per inch, substantially enhancing fine details and overall image sharpness. Sequin Digital Screening can produce high-quality images without large volumes of data. Sequin produces a sharper apparent resolution and finer color detail. Other advantages include an expansion of the number of gray levels up to 256 per separation and the elimination of moiré.

The Booklet Maker is an automatic finishing option that enables the E-Print to produce fully finished booklets of up to 80 pages. Printing, collating, folding, and stitching are carried out in a single step without intervention.

The ability to change the image for every impression gives the E-Print 1000+ exceptional abilities for individual personalization of printed materials. The E-Print 1000+ can split a page into areas so that only the variable information is modified on the cylinder to "personalize" a page. This keeps data transfers to a minimum and press speeds to a maximum.

## Indigo E-Print Offline Workstation

The Indigo E-RIP is an offline workstation for job processing for its E-Print 1000 Digital Offset Color press. E-RIP will handle offline processing tasks performed currently on the E-Print 1000's integrated workstation. These include job output, layout, parameter definition, processing, and all data preparation for printing.

Job processing times naturally vary depending on job layout, complexity, and picture content; therefore the ability to balance data preparation and flow is a valuable benefit in the digital print shop. Offline job input and processing helps balance the workflow for complex jobs, or very short runs, where the job preparation may sometimes be longer than the printing.

From a production management standpoint, operators can be preparing one job offline from the press while another is being ripped or printing. E-RIP is particularly useful for workflow involving complicated jobs, very short runs, and jobs requiring multiple changes. E-RIP also offers workflow and productivity enhancements by processing PostScript jobs offline, allowing the E-Print operator to concentrate on printing.

The offline RIP and workstation comprises a SUN Microsystems platform equipped with high-performance data storage and LAN communication hardware, running E-Print 1000+ and Adobe RIP software. The product comes as a complete package including all system integration, support, and operator training.

## Personalization

As discussed in the chapters on personalization, the personalization ability of on-demand digital printing on the fly has been one of short-run digital printing's most alluring promises, but one that has taken a good deal of technological development to implement

Yours Truly™, Indigo's trademarked personalization option, not only makes real-time personalization an affordable reality but is being used increasingly for a variety of precision-targeted full-color products. American Digital Services, Inc., a digital imaging company focusing exclusively on short-run four-color printing, is using the E-Print 1000+ to produce complex and sophisticated personalization applications: 1,316 different 8.5×11-in. (216×279-mm) customized four-color pages for an internal sales and marketing organization. Each page had approximately fourteen different changes in text or graphics or both, with all of the changes driven by the customer's critical database files.

The job was produced at full press speed, with each page's customization made on the fly. Take a minute and try to figure out how you could do something like that in an affordable time frame on a conventional press that requires film separations and plates.

## E-Print 1000+

Indigo has launched the feature-rich E-Print 1000+, the first new version of its Digital Offset Color press in three years. The company has upgraded its entire worldwide installed based of E-Print 1000s to the new E-Print 1000+ machines. All new installations will be E-Print 1000+ presses. Indigo set the list price of the new, unbundled version of the E-Print 1000+ at $379,000. It significantly reduces the price-point of entry for high-quality digital short-run color. The unbundled version comes with the ability to print PostScript files, monochrome personalization, and 2,000 letter-size, four-color pages per hour with a resolution of 800 dpi.

The company has introduced an enhanced line of imaging products including newly designed ink cans that increase yield by 30% and a new Photo Imaging Plate. The improved performance of these products, combined with new lower prices, result in a reduction of 50% in the list price of Indigo's consumables cost per 8.5×11-in. (216×279-mm) page. At the same time, they substantially improved the E-Print's functionality and performance.

Indigo also announced the forthcoming availability of an EFI Fiery front end for the E-Print 1000+ digital press that makes production-color printing as easy and as accessible as color copying with higher quality and lower cost per page. The Fiery controller will be the newest of several options for the E-Print 1000+ press. Others currently include full-color personalization, high-definition imaging, automatic duplexing, and expanded memory.

The actions announced were designed to address two major company initiatives: improving current Indigo users' experiences with their machines, and helping the company and its customers capitalize on the growing opportunities in short-run digital color.

Indigo also announced a simplified, lower list price structure for its entire line of imaging products for the world's most widely used full-color digital press. The dual impact of Indigo's lowered list pricing, coupled with release of the company's higher performance imaging products, means immediate, major productivity gains for Indigo digital press owners. The result is the digital printing industry's lowest overall costs per impression and wider margins to enhance the profitability of short-run, on-demand printing operations.

Following successful beta testing in the United States and Canada, Indigo has released the newest version of its Photo Imaging Plate (PIP). The product of Indigo's ongoing research and development program in its Israel-based laboratories, the new

PIP incorporates the company's latest advances in electronic digital imaging technology. It provides improved on-press performance, higher quality results, and an appreciable gain in operating productivity. The Indigo PIP is capable of printing a new billion-dot image impression every 0.9 sec. and is at the center of the E-Print 1000+'s ability to print a completely new full-color image with each revolution of the plate cylinder.

In another important product performance advance, Indigo has begun distribution of its new Indigo ElectroInk canister and related ink delivery system. The larger ElectroInk canister delivers 30% more 8.5×11-in. (216×279-mm) impressions than the prior existing version. This larger can also employs a unique, efficient piston that consistently drives virtually all of the contents of the ink cans into the liquid ink reservoirs of the E-Print 1000+ digital press.

In announcing the new lower list pricing structure for these and other Indigo imaging products, the company noted that they now reflect actual discounted transaction prices that Indigo owners have been receiving. For example, the new Indigo list price for a canister of ElectroInk, before earned discounts, is $39.58, compared with the previous list price, before discounts, of $115 per can.

Indigo realigned its list pricing to more clearly reflect what customers have in fact been paying all along with Indigo's discount structure. Now with higher-yield imaging products entering the distribution pipeline at the current pricing levels, Indigo E-Print 1000 digital press owners will have more flexibility than ever to enhance their operating margins.

Indigo's Digital Offset Color printing technology combines the quality of offset printing with the flexibility, economy, and ease of use of electronic printing. The company's unique ElectroInk liquid ink technology produces brilliant magazine-quality glossy color images. This fully electronic imaging process produces just-in-time, on-demand printing without films, plates, proofs, or traditional makeready.

Indigo provided us with current pricing levels for all E-Print 1000+ consumables, and clarified various points for us. See the following table for yield and cost of consumables.

## Yield and Cost of Consumables for Indigo

| Consumable | Unit | 8-15-96 Price | 11×17 Yield | 8.5×11 Yield @ 30% Coverage | 8.5×11 Yield @ 60% Coverage | 8.5×11 Yield @ 45% Coverage | Cost 8.5×11, 4C, 1S @ 45% Coverage | Cost 8.5×11, 4C 2S @ 45% Coverage |
|---|---|---|---|---|---|---|---|---|
| Blanket | Blanket Cartridge | $550 | 60,000 | 120,000 | 60,000 | 90,000 | $0.006 | $0.012 |
| PIP (Plate) | Plate Cartridge | $380 | 40,000 | 80,000 | 40,000 | 60,000 | $0.006 | $0.009 |
| Imaging Oil | Bottle | $8 | 5,000 | 10,000 | 5,000 | 7,500 | $0.004 | $0.009 |
| ElectroInk | Canister | $32 | 3,900 | 7,800 | 3,900 | 5,850 | $0.022 | $0.044 |
| | | | | | | | $0.039 | $0.077 |

**Note:** One impression equals one color over an 11×17-in. (279×432-mm) sheet.

The pricing for ElectroInk, for instance, has always provided discounts for quantity and payment. Thus most users were not paying the list price of $115 per can; most were receiving a 45% discount for certain volumes, which made the price $63.25. An incentive program for prompt payment offered two cans for the price of one, which brought the price down to $31.63. To simplify the entire process, Indigo set a street list price of $39.58 to reflect what most users were already paying and then offered an incentive for prompter payment of 20%, which brings the street price to $31.67.

At the same time, Indigo increased the yield from plates and ink. A different approach is used to force the ink out of the can. In addition its yield also increased for other more technical reasons. Indigo is using 3,900 11×17-in. (279×432-mm) sheets at 30% coverage for the yield, but this appears to be conservative.

The pricing presented to us is impressive and changes some of the assumptions we have made in our costing model. Indigo arrived at $0.03 for four colors one side, at 30% coverage, which would be $0.06 for both sides, only counting the consumables.

Our only area of difference is in the coverage, a subject we continue to grapple with. Thus, we have extrapolated from 11×17 to 8.5×11 sheets and from 30% to 60% coverage. Then we took an average of 30% and 60% to better reflect what industry averages may look like. Thus we arrived at $0.04 to $0.08, which is still very good.

## Label Market Solutions

Indigo announced that it is offering a digital printing solution for the label-printing industry built around the Omnius Digital Offset Color press, a new white ink, and the broadest selection of substrates available in the industry. During Labelexpo 96, the Omnius digital press was demonstrated at Indigo's exhibit and also at the 3M, MACtac, and Fasson exhibits.

The Omnius web-fed digital press was introduced in early 1995 and has been successfully installed in the operations of several major customers around the world. During the past year Indigo has formed partnerships with leading label substrate manufacturers, with the result that 39 different substrates are now available optimized for Indigo's proprietary ElectroInk liquid-ink technology.

One of Indigo's latest substrate announcements was in conjunction with 3M which extended the current portfolio of 3M's D.I.G.I.S.T.A.R. products, originally launched in August 1995, optimized for the Omnius.

Indigo has also introduced white ElectroInk for use as full-background or spot color printing on transparent flexible packaging and label substrates. The white ink release follows successful research and customer beta-site testing.

Label printing is already done primarily in short runs, and about 80% of label runs are under 100,000. The Omnius is designed to handle these runs from a few dozen variable-image prototypes to full production runs. Label pressruns are getting shorter and turnaround times faster as marketers are customizing packaging by geography or market segment, making fast changeovers in response to consumer attitudes, buying patterns and new product attributes, conducting target marketing, and test-marketing products on an ever-increasing basis all the while trying to eliminate expensive and potentially obsolescent packaging inventories.

The Omnius is a fully digital press that prints in full color directly from electronic files, including Adobe PostScript and Scitex HandShake. It works with industry-standard application software to offer variable color, images, and text, label by label.

Substrates for the Omnius are offered by most industry leaders, including 3M, Fasson, MACtac, Mobil, and PPG. Indigo has developmental alliances with several leading companies in the electronic imaging, communications and printing industries, including Adobe Systems, EFI, 3M, Champion International, Mobil, PPG, MACtac, S.D. Warren, and Fasson.

## Customer Feedback

When first introduced, Indigo received some harsh criticisms. First there was the "ignore the man behind the curtain" objections when hundreds of show attendees could not see the press because it was draped. Next was the requirement to buy two presses because "we can't really keep two presses running." And finally, there were the erasability problems.

However after a few years on the market and some significant engineering changes, the majority of customers conclude that the Indigo press has some distinct advantages. The unique ElectroInk technology enables brilliant magazine-quality glossy color images. A fully electronic imaging process produces just-in-time, on-demand printing without films, plates, proofs, or makeready.

The E-Print 1000 brings the printing industry the best of both worlds. Printing directly from digital data and accepting industry standard formats like PostScript and Scitex, the E-Print 1000+ can operate as a stand-alone unit or network with prepress and desktop publishing systems.

The press accepts a broad range of coated paper stocks as well as uncoated paper up to 12×18 in. (305×457 mm) in size. At a process speed of 120 ft./min. (37 m/min.), it can print 2,000 full-color (or 8,000 single-color) letter-size images (two-up) per hour. With a standard printing resolution of 800 dpi, the E-Print 1000+'s proprietary image enhancement technology achieves outstanding, crisp, brilliant images, both for text and color graphics.

Lastly, the company has also lowered by more than half the list price of its imaging products.

# Chapter 31

# Scitex Digital Presses

We first heard about the Scitex inkjet press at the Scitex Users Group in August 1993, in Nashville, Tenn. It was quite a shock to everyone, considering that Scitex had acquired Kodak's Dayton inkjet facility only a few short months earlier.

In June 1993 Scitex acquired Kodak's Dayton Inkjet Operations and renamed it Scitex Digital Printing Inc. (SDP) — Kodak had acquired the technology from Mead. Over the years, this facility has become recognized as a leader in high-speed inkjet technology with over 200 patents. At the time, the equipment printed black-and-white pages at 240 dpi, 1,000 ft./min. (305 m/min.) and was used to personalize direct mail, sales promotions, magazines, as well as for database marketing and addressing.

At the press conference in Nashville, Yoav Chelouche, corporate vice president of Marketing and Business Development, drew a chart of inkjet technologies with a speed axis and a quality axis. Close to the end of the speed axis he placed the word "Dayton" and close to the end of the quality axis he placed the word "IRIS," the Scitex-owned company that makes high-quality inkjet proofers.

In the middle of the graph he wrote "Product Heaven." The point is that there is a market for both high-quality and very fast inkjet products, however if you could combine both technologies there may be an even larger market.

In the ensuing discussions, it became clear that inkjet technology, with some improvements, could be used as a printing press. Chelouche described the different pieces that would need to be brought together to make a "Scitex Press."

There are front-end technologies that Scitex already possesses, core technologies such as high-speed, high-quality printing that Scitex could refine, and the mechanics of paper handling that Scitex could partner with another company to bring this product to market. The take-home message was the possibility of a Scitex Press. A prototype of this unit was first shown at an open house at the Dayton plant on March 8, 1995 and then shown again at Drupa.

For the people that follow inkjet technology, Scitex Digital Printing, Inc. division is a leader in the development, manufacturing, and sales of high-speed inkjet printing. The technology is based on over 170 registered patents and 30 years of experience. Like Iris and Leaf, Scitex Digital Printing is a wholly owned subsidiary of Scitex Corporation Ltd.

The existing Scitex inkjet technologies are used in the commercial printing, business forms, direct mail, direct marketing, promotional graphics, catalog, magazine, lottery, and gaming industries as well as in-plant printing and mailing operations.

In 1995 Scitex Digital Printing revenues exceeded $100 million, up almost 20% over 1994, with strong operating margins. The growth was achieved primarily through a significant broadening of its customer base and the development of new geographical areas including Japan, Asia/Pacific, and Europe.

The outlook for SDP is excellent, as evidenced by its announcement of a $35 million order from a major manufacturer of business forms systems in Japan. They expect the strong growth to continue through the expansion of traditional markets, as well as through development of new areas, such as the data center, forms printer, and tag and label markets. SDP is also exploring growth opportunities in the packaging, newspaper, and demand publishing industries.

## Scitex 3600 High-Speed Printing System

There are three products in the Scitex 3000 Digital Printing division: Scitex 3600, Scitex 3000, and Scitex 3500. The 3600 and 3500 can print variable, 240-dpi output at 1,000 ft./min. (305 m/min.) and 500 ft./min. (152 m/min.), respectively. The 3000 prints at 120 dpi.

The 3600 can be configured with 2–8 print stations for full-page printing of 240-dpi text, bitmapped graphics, and bar codes. Output from multiple devices can be combined to produce images across 34 in. (864 mm).

When getting personal means full-page, 100% variable data imaging anywhere across a 40-in. (1016-mm) web, the new Scitex 3600 does it faster than any other system in the world. You can actually change 100% of the data from one piece to the next, "on the fly," at speeds of up to 1000 ft./min. (305 m/min.). This means faster throughput and improved productivity for high-volume personalized direct mail, sweepstakes, lottery tickets, business forms, financial statements, and many other variable printing applications. It also means full-page images with letter-quality text, bitmapped graphics, and bar codes.

Exclusive use of multiple, independent print modules provides complete creative freedom. Two to eight 4.27-in. (108-mm) printheads let you position variable or fixed data anywhere across the full web width or even duplex. You can do spot printing, spot-color printing, two-up printing, or even generate graphic overlays. For additional flexibility, both systems can store up to 200 fonts on-line per job. Any of these fonts can be used on any line anywhere on the page, and even rotated at full press speed.

Beyond these standard capabilities, the 3600 High-Speed Printing System offers a number of performance enhancements. These include a 32-MB memory upgrade allowing you to handle the most demanding graphics-oriented applications, a second RIP, and the ability to image with up to eight printheads as well as duplex and synchronize two high-speed systems.

## Other 3000 Series

The 3500 prints 240-dpi text and bitmapped graphics at 500 ft./min. (152 m/min.). This product is targeted at shorter and more-targeted run lengths and for use with in-line finishing equipment such as gluing, perforating, cutting, and plow folding stations. The 3000 prints 120-dpi bitmapped graphics and text and can print rotated text or variable information. The maximum size is 13.3 in. (338 mm) and maximum speed is 100 ft./min. (30 m/min.).

These web-fed, inkjet devices are designed for printing personalized pieces in volume with quick turnaround. The technology allows for 100% variability, but may be best used at imaging variable components within static page layouts — in other words, personalization and customization.

As described in the chapter on database marketing, the products or services for printers capable of variable printing are mostly direct mail such as personalized letters and documents. Although not a demand or digital printing product, another application is lottery tickets. It may be a customized product, however, if the ticket is numbered.

According to the Scitex officials, the 3000 Series systems produces 85–90% of all scratch-off gaming tickets in the country. Another application for the 3000 series is forms printing. This too is an application that may use variable data, with some of the form filled in.

These devices can print either fixed or variable information on each page. Another interesting usage of this technology is when configured "in-line" with a web press. Used in that configuration the offset press would print the fixed information, which could be color, while the personalization would be one color (i.e., black).

## Scitex 5240/5120 Printers

Scitex 5240/5120 printers employ patented single-row inkjet technology to ensure high-resolution, superior edge definition, precise alignment of character elements, and consistent spacing between lines. Theses characteristics make them ideally suited for inkjet printing applications that demand exceptional precision and image quality (e.g., direct mail, business forms, gaming tickets, bar coding applications, and greeting cards).

Image areas are 1.067×22 in. (27×559 mm), with standard resolution of 240×240 dpi for the 5240 and 120×120 dpi for the 5120. Resolutions of 240×480 dpi (5240) and 120×240 (5120) are also obtainable. Printheads can print in any orientation and are self-cleaning, which virtually eliminates traditional printhead maintenance.

Printers are available with either a standard flexible, 12-ft. (3.6-m) umbilical cord or an optional 24-ft. (7.3-m) flexible umbilical and a comprehensive set of accessories that make them adaptable to just about any graphics application requirements.

## The Prototype

The prototype is targeted at the direct mail industry, to print customized pieces; the business forms/tags and label printing market, to accommodate very short runs and faster turnarounds; lottery printing market, using color images, text, play symbols, and sequential numbers; catalog market, for customized catalogs; and customized periodicals.

As demonstrated in the Dayton facility, the prototype printed on an 8.5-in. (216-mm) web with black text printed on the left and full color printed on the right at a speed of 200 ft./min. (60 m/min.).

The printheads house a linear array of inkjet nozzles that can be charged or not. Charged droplets return to the ink containers for reuse. A patented, single row of jets is provided to enhance edge definition, registration, and positioning. There are four heads: one each for cyan, magenta, yellow, and black ink.

A continuous sequence of pages can be printed at a fast speed. The speed of 200 ft./min. is slow when compared to the other Scitex monochromatic counterparts, but not slow when compared to other digital color presses.

The Indigo E-Print 1000, for example, prints 1,000 A3 four-color impressions per hour in four colors, each of which is slightly less than 2 ft.(610 mm), for a maximum printing speed of less than 2,000 ft./hr. (610 m/hr.) for four color or 33 ft./min. (10 m/min.) for color (133 for single-color).

The prototype only printed one side (simplex) and was running at 200 ft./min. The company says the commercial version would be a duplex model with an image area to 18×34 in. (457×864 mm) and quality comparable to 800×800 dpi, which ironically is the original Indigo resolution.

## Software

The advantage of this technology is that it is very fast, and using it has become easier with new software called Begin. The new Begin software uses QuarkXPress to create page templates for output on the 3000. Previously done with mainframe-based software using a user-unfriendly command language, the use of QuarkXPress makes page setup much easier. The new software controls three basic functions: setting up the position of the printheads on the web, laying out the pages, and merging the variable data.

Using the Begin software, the design portions of the job are done with QuarkXPress. In this workflow, the only portion of the work that requires programming-type skills is the data-merging step. Furthermore, consistent with the general trends today, the mainframes are no longer required. The database portion can be done on a Sun workstation.

The database merging functions are done on the Sun workstation using various tools. First, the QuarkXPress layouts are loaded onto the Sun workstation where the database, typically a nine-track tape, is input. The variable portion of the information is combined with the fixed data, and a tape is created for the entire job. This tape is then used to drive the printer.

The price of the Begin software is about $40,000, which does not include the Sun workstation. Even if the combined price was $100,000, it would be a small price considering that the Scitex Digital Press ranges from $1–3 million.

## Issues

The disadvantages of the prototype as it exists today is that it does not print PostScript, is limited to 240 dpi as opposed to the 300–600 dpi of other devices, and personalization is limited to certain areas.

The quality of the type and images will be an issue. An initial market will probably involve newspapers, and we could see an inkjet attachment at the end of a newspaper press. We also see more than black printing. The newspaper could feature a house in a real estate section, and each house could be selected based on the zoned edition of the newspaper.

The only areas along the web that can be customized are those that fall under a printhead. The 240-dpi printheads are approximately 4.3 in. (109 mm) wide (1,024 pixels), and several heads can be combined into a larger print area. Depending on the number of "bells and whistles," you can have one or more printheads, extra printheads for spot color, or two print units with a turn bar between them so that you can print on both sides of the web.

The exact number and positioning of the printheads has to be configured with the "Web Layout" module, which is a Quark XTension. The module is interactive and shades the portion of the web or pages that are printable.

After the page layout is complete, the software allows you to position it on the web and will inform you if you have placed them in an unprintable area.

The Scitex Digital Press and the Begin software makes it possible to use the 3000 Series printers for a few applications. But most applications will be hampered by one of the system's limitations. The lack of PostScript, the relatively low resolution, and the limitations in inkjet coverage limit the applications. But these problems can be addressed by tomorrow's technology. Of course, the quality will improve and the cost will decrease.

Although the product may not be ready for commercial sales for 18–24 months, the technology certainly warrants a closer look.

## Goals

To achieve the goal of increased size as well as the goal of improved quality, several advances are required. Among those that have already been achieved include smaller printhead nozzles and droplet sizes, a variable number of ink droplets per pixel (for greater resolution), new process inks, and a switch from continuous flow to "on-demand."

One of the first redesigns was on the orifice of the printheads. The new printhead nozzles are about one half the diameter of the conventional heads. The resolution of 240 dpi has been maintained in the horizontal direction, but the resolution in the direction of the web movement has been increased to 1,680.

The smaller nozzle and increased resolution was the result of the collaboration with the engineers from Iris. They pioneered the use of "variable dot size" in which the number of ink droplets per pixel location can be varied from 0 to 32 droplets per pixel. The result is that the Iris inkjet 300×300-dpi resolution has an apparent resolution of 1,024×1,024 dpi.

The new prototype is capable of spraying seven individual droplets per pixel location, or $7 \times 240 = 1,680$ dpi. According to Scitex the 240×1,680 resolution is comparable to between 600- or 800-dpi apparent resolution.

Among the hardware advancements still required are wider printheads, duplexing, new hardware for faster data handling, color correction software, and a new printing system design.

## Spontane

At Drupa 1995 Scitex also introduced the Spontane color printer based on Xerox copier technology. It is covered in detail in the chapter on "Copier/Printers."

# Chapter 32

## Other Systems

Electrophotography is not alone in applicable technology for on-demand approaches. We have already seen inkjet being applied. In our chapter on "Futures," we discuss some systems now in development. There has been a continuing search for alternatives to electrophotographic printing because of the following deficiencies:

- Charge decay between the time of charging the photoconductor, exposing the image, and toning can affect image density and tone reproduction. The amount of toner transferred to the image is dependent on the exact voltage of the charge at the time of toner transfer.
- Toner chemistry can cause variations in batches of the same toner and high cost for formulating special toners.
- In liquid toner systems the isopar used to disperse the toner is a volatile organic compound (VOC), which may require venting and is subject to environmental regulations.

### Stencil Duplicators

Stencil duplicators as in many other electronic publishing, digital advances are giving new life to old methods. Remember those purple ink worksheets? Well, that's the past of duplicators. Imagine now producing 3,000 copies with different colors in less than two hours, or producing other kinds of jobs even faster. This is possible nowadays because of new stencil duplicating technology. This technology uses a computer controller with digital imaging to fill a specialized printing market.

**History.** Stenciling is the reproduction process known under the generic term of mimeographing, invented at the turn of the century by Thomas Alva Edison and Albert Blake Dick. In this process there are two steps: producing the stencil by typing it on an impact typewriter and mounting the stencil on a "press" and running sheets of paper through the duplicator.

In 1955, RISOKayaku of Japan was established by its founder and current president, Noboru Hayama. In the beginning RISO concentrated the majority of its efforts on the research and development of emulsion inks. When the inks started to became popular, the group started to investigate stencils and duplicators. Throughout the 1960s and 1970s the research department of the company only focused on the improvement of the inks, papers, and films. Around the 1970s, the company started to produce a high-speed thermal master. This discovery led the company to the development of the first stencil duplicating system.

After the development of the machine, RISO introduced a new technology called "risography" during the 1980s. In 1986, RISO was established in the United States. From this location, RISO did a great effort with the marketing and the sales of these machines in both North America and South America.

Nowadays RISO is the only company that produces its own parts. In this way, RISO guarantees the performance and the quality of the equipment it sells. RISO holds two patents for the inks and the master. The use of RISO's inks guarantees consistent print quality in any color or combination of colors. The ultra-thin master material allows the printing equipment to get an extremely high print resolution, resulting in high-quality output.

**Technology.** Stencil duplicators are not designed to completely replace copiers or offset presses but to carry out jobs requiring twenty or more copies per original and also bridge the gap between the copier and an offset press. Stencil duplicators offer fast on-demand color designing and printing for very reduced costs. For documents that typically range from one to a few pages in length and do not require very high image quality, a stencil duplicator will handle the work incredibly.

From the outside, digital duplicators look like photocopiers with a control panel from which the number of copies are selected. But that's where the similarity ends, because digital duplicators uses other technologies totally different from the normal photocopiers. For example, digital duplicators do not use toner; they print with ink using a special master, eliminating the extreme heat and pressure required in normal photocopiers.

In a sort of electronic silk-screening process, today's duplicators receive digital signals from computers that cause a thermal head to burn perforations in a photosensitive master. The master, a few microns thick and made from polyester resin film bounded to thin fibrous paper, is wrapped around an ink drum. Inside the drum a squeegee device presses ink through the perforations of the stencil. Pickup rollers guide paper past the drum, and ink is transferred onto the paper. Colors are added by simply changing the ink drums and running copies back through the duplicator.

Although stencil duplicating technology rivals the short-run market printed with offset, the technology has some limitations. It can not print continuous-tone images, and it can not print on coated stock.

Trapping is one of the biggest limitations in this technology. Misregistration can be controlled by keeping the machine as clean as possible. Some companies said that by keeping their machine and the environment dust-free, registration can be controlled within one inch. RISO say that these machines can hold registration to an eight of an inch, however this can be improved by keeping the machines in good shape and give them good maintenance.

Although these machines work very well as stand-alone devices by duplicating originals with internal scanners, sending documents directly from a Mac or PC without using the scanner results in better quality. Multicolor documents must be processed through an illustration or page layout program, so the duplicator will get the data for each spot-color separately.

Duplicators come in different configurations. You can decide from the following variables: amount of RAM, the number of built-in fonts, and the choice of collators, stackers, or input trays you want. A top-of-the-line duplicator can be purchased for less than $35,000. Another feature of this technology is that it is environmentally friendly and energy efficient. The ink that is used is water-based, and spills are very easy to clean up. As for energy consumption, to give an example, we can compare a duplicator with a high-end xerographic copier, and observe that 50,000 copies on the copier cost $86.34 in electricity charges while the same job cost $1.75 on a stencil duplicator.

Five companies currently sell digital duplicators in the United States: A.B Dick, Gestetner, Ricoh, Riso, and Standard Duplicating Machines. The companies that market their machines most actively are Gestetner, Riso, and Ricoh. Riso centers its technology on stencil duplicating and coined the term risography, This company has the largest market share (55.5%), followed by Gestetner with 16.3% and Ricoh with 5.8%. (Ricoh produces the print engine for Gestetner, and Gestetner produces the computer controller for Ricoh.)

Risograph's technologies combines digital facsimile scanning, copier electronics, and a self-contained, fully automated high-speed printing system. Risographs can be linked to PC or Apple Macintosh computers via Riso's interface. In this way, you can output directly from the computer with no intermediate hard copy. The Riso RC series combines desktop publishing with duplicating technology. It will allow forms to be scanned into the computer and stored as TIFF files. The forms then can be called up and printed on the Risograph at 130 pages per minute on a variety of substrates with a choice of twelve spot colors. With these duplicators, colors are added by changing ink drums and running copies back through. Documents are sent directly from computers to the thermal head. The duplicator is a PostScript device, providing emulation so the computer handles it like a laser printer.

Recently Riso introduced its new model, the Super Risograph SR 7200. It incorporates several design improvements. The new polyester drum, rather than the metal drum of existing units, provides better solid blacks. A new system of feed gripper and an impression cylinder provide better registration. The SR 7200 is an A3 machine. There is a scanner for originals and a PostScript input option. In this machine, the master (which is created in 35 seconds) is automatically fed from a roll of 200 and automatically compressed and deposited in a used master bin after printing.

Pitney Bowes Copier Systems has a 92-cpm C950 Fast Track Copier Duplicator. The document processor consists of Reversing Automatic Document Feeder (RADF), Semi Automatic Document Feeder (SADF), and a Computer Forms Feeder (CFF). The RADF operates at full machine speeds and feeds up to 100 originals ranging in size from 5.5×8.5 in. (140×216 mm) to 11×17 in. (279×432 mm). Users can feed odd-sized and fragile originals for automatic two-sided copying.

## AM Electropress

The AM Electropress was displayed at Drupa 90 with a videotape of an installation in Sweden. It is an expensive two-unit webfed press that uses early Indigo electrophotographic technology. It has two cylinders coated with organic photoconductors and liquid toners (ElectroInk) to produce the image, in one color on both sides of the paper or two colors on one side. The press prints variable information at speeds up to 300 ft./min. (91 m/min.). The press is capable of changing about 50% of the information per unit from impression to impression. There are about 30 presses installed worldwide, and they are used primarily for direct mail applications.

## Magnetography

Magnetographic presses were shown at IPEX 93 by Nipson Printing Systems (Bull Groupe, Belfort Cedex, France) and GMC Digital Systems (St. Gallen, Switzerland). Magnetography is a nonimpact print technology like ion deposition (Delphax) except that a magnetic drum is used, a magnetic charge is produced on the drum by an infinitely variable computer-generated image, and a monocomponent magnetic toner is used. Spot colors can be used but not process colors as the toners are dark and opaque. The systems are used for printing business forms, direct mail, lottery tickets, numbering, tags, labels, and bar codes. The Bull engine is used almost exclusively for these systems.

Nipson, which is owned by the Bull Group, has its Varypress M700. This device has speeds from 60 to 345 ft./min. (18–105 m/min.), has a resolution of 240×240 dpi with web width capacity of 20.5 in. (521 mm) and a print width of 18 in. (457 mm). Nipson also had the M420 and M450 Varypresses on display. These devices are limited to a 14-in. (356-mm) print width.

GMC uses the Bull engine and has shown two models of systems:
- **Speed Robot,** with image width of 14 in. (356 mm) and a maximum printing speed of 460 ft./min. (140 m/min.)
- **Speed Robot Model 520/18,** with image width of 18.2 in. (462 mm) and maximum printing speeds from 41–62 ft./min. (14–20 m/min.)

## Ionography

This process is also known as ion deposition or electron-charge deposition printing. The process was developed by Dennison Manufacturing Co. and is called Delphax. The image is produced by negative charges from an electron cartridge onto a heated dielectric surface of aluminum oxide using a special magnetic toner. It is used only for single or spot-color printing because the pressure of image transfer and cold-fusion fixing of the toner can distort the substrate. Over 3000 systems are in use for volume and variable printing of invoices, reports, manuals, forms, letters, proposals, and specialty printing of tags, tickets, and checks. A four-color system is claimed to be in development.

EBI process by Digital Print (Ft. Worth, Tex.) is an improved ion-deposition type (Delphax-like) digital printing system rather than an electron-beam imaging system like the IGI system by Image Graphics, Inc. (Shelton, Conn.), which is used to produce large quantities of high-resolution (over 8000 dpi) 70-mm films for high-volume printing. The Digital Print electron beam process has an electron deposition print cartridge that deposits a charged image on a dielectric cylinder which is toned, transferred, and fixed to a substrate with pressure. Printing speed is up to 225 ft./min. (slow) at a user-selectable resolution of 240 or 300 dpi.

# Chapter 33

# Futures

There are a number of development projects in the works focused on the application of new technologies for on-demand printing. Some have been publicized, while others maintain a cloak of secrecy. We have placed those that we have information about in this section. Please keep in mind that these are developmental systems, and it may be years before they become products.

## XMX and Field Effect Imaging

According to Mike Bruno, well-known industry consultant, this process — invented by Dr. Manfred R. Kuehnle and developed at XMX Corporation (Billerica, MA) — uses three advanced materials to produce high-speed, high-quality color printing composed of variable-density pixels with resolutions of 500 dpi.

The new materials are X1, which is a thin-film dielectric ultra-hard writing surface onto which electrical charges are deposited by an M-Tunnel write-head to generate powerful electrostatic fields across the thin-film insulator, which fields cause the pickup of X2 "ink bites" whose thickness is proportional to the strength of the fields. The bites of ink, in pixel size, are carried by the X1 surface to a print position where the ink is transferred totally to the print substrate, which can be either paper, plastic, cloth, or metal.

According to XMX, the functional principles differ completely from those employed in classical or conventional imaging technologies and thus overcome the limitations in speed, quality, and predictability that now apply to the use of these technologies. The XMX printing system can print fully variable-color imaging on demand at speeds in excess of 1 m$^2$/sec., which equals a 2.4 gigahertz bandwidth per color.

No published material is available on this system. It has been said that the system can print on 20- to 40-in. (508- to 1016-mm) sheets, duplexed. A model of the machine described at a press conference in 1991 in Canada appeared to be about 20 ft. (6.1 m) long, perhaps accommodating color transfer units for four or more colors on one side and then four or more colors on the second side.

## MAN-Roland Digital Change-Over Web

MAN-Roland, a German press maker, showed its DICOweb (Digital Change-Over) technology at Drupa 1995. The technology involves writing to four printing cylinders directly and then erasing them and rewriting, etc. for short-run printing. The process uses a ribbon of hydrophilic material that is brought into contact with the cylinder, which is then laser-imaged from the back. The thermally activated areas adhere to the cylinder and then become ink-receptive.

A laser made by Creo burns 11-micron spots. The sleeve that carries the image requires a certain degree of roughness. After the image has been burned onto the sleeve, the image is remelted with hot air to bond the resin image more firmly. A letter-sized page would take 3 min. and a tabloid-sized area would take 5 min. to image and fix.

Once imaged, the press runs at 45 miles an hour — 20 meters per second. Future machines would come in 18- to 26-in. (457- to 660-mm) widths. If you do the arithmetic, a tabloid sheet in a quantity of 500 would take about a second to run. The imaging of the sleeve for single-sided printing would take about 13.5 min. — the goal is to get it to 5 min. or less. The sleeves may be able to be re-imaged 50 times. Since even a molecule of the image substance would create problems, cleaning is essential.

 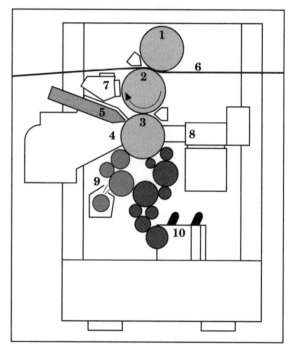

Above left is an illustration of the removable sleeve that carries the image on the DICOweb offset press. The schematic at right shows the impression cylinder (1), the blanket cylinder (2), the imaging cylinder (3), the cleaning unit (4), fixing unit (5), paper path (6), the blanket washer (7), laser imaging unit (8), dampening unit (9), and inking unit (10).

The image is erased by a device similar to a blanket washer with a special solution. Resolution is now at 2,400 dpi, and there is no idea about run length — jobs of 500 to 5,000 copies are predicted — or how many times the cylinder surface can be rewritten — they think it could be 20 times but it certainly will be based on paper type and run length.

The 64-channel laser now takes 5 min. to make an 11×17-in. (279×432-mm) image area but the goal is 2 min. The speed is 2 m/sec. (78 in./sec.), but the goal is 50 m/sec. (164 ft./sec.).

Quality of the printed output was poor, but the system was shown as a future technology, not as a product ready for market. It could be a few years before we see it in real use. A major advantage is that it runs on a system created for web presses with drying, chill rolls, and cutting online. The concept is quite valid and would be classified as a hybrid printing press — one that makes the image carrier on press.

This system only prints on one side at present. A variation involves creating the image for the second side off-line on a thin film and then transferring the image. This can only be one color, however. MAN Roland has been careful not to call the cylinder and/or sleeve a plate. Presstek holds a systems patent that covers making a plate on a printing press.

A variation of the MAN Roland technology works with gravure printing as well. A metal or ceramic cylinder has all gravure cells pre-engraved and full depth. A dark, waxy polymer fills every cell. A laser removes the cell contents to selected depths by vaporization. The image is transferred to a blanket for printing. Runs of 500 to a few thousand are predicted and the cylinder can be used for about 20 jobs or so.

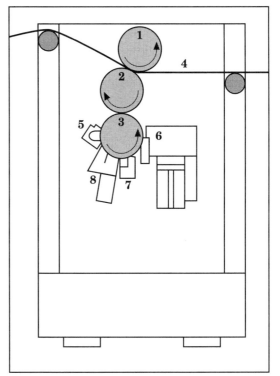

The image cylinder (3), the blanket cylinder (2), and the impression cylinder (1) of the DICOweb gravure technology are shown at left. The filling unit fills all cells with the waxy substance (7), which are imaged by the laser (6). Ink is removed with a doctor blade (5), and the cylinder is cleaned with an erasing unit (8).

## Mitsubishi

Mitsubishi Heavy Industries has introduced its Digital Press that uses a web of paper like the Xeikon engine, the printer is rated at 800 dpi and 72 11×17-in. (279×432-mm) pages per minute, single-sided. The Xeikon does 72 letter-sized pages per minute, duplexed. The web of paper is 12.32 in. (313 mm) wide. The pages are pre-rasterized and stored in the system memory, which can hold 200 color pages or 800 monochrome pages.

There are still many questions to be answered, but this could be among the second wave of digital presses, bringing on-demand color printing to a larger audience.

## Toray

Toray showed its CP-3120 at Drupa for short-run color printing. It produces four letter-size four-color pages per minute, or half that for tabloid-sized sheets. The system is about the size of a typical color copier and has a resolution of 800 dpi. It provides duplexing as an option. Like the Indigo E-Print, it uses liquid toner, and a Harlequin RIP is used. A price tag of $100,000 has been announced, even though copy quality may not be up to its competition.

## Elcography

There is a new printing technology in development that uses an electric current to coagulate ink on the drum of a web press. The technology is known as "elcography" and is under development by Elcorsy Inc. in conjunction with Toyo Ink and Intex Inc., a Japanese financial firm. As we go to press, it has only been demonstrated once on an 11-in. (279-mm) wide digital web press, but according to rumors, subsequent demonstrations will be on 24- and 30-in. (610- and 762-mm) web presses.

As demonstrated, there is a printing unit for each color, and the prototype output speed was 20 ft./min. (6.1 m/min.). The maximum speed using this technology is 3,000 ft./min. (914 m/min.). According to Elcorsy the 24-in. (610-mm) press running at 200 ft./min. (61 m/min.) will output 38,000 full-color tabloid newspapers pages per hour.

Elcography was invented in 1971 and is based on conductive polymeric inks that coagulate rapidly as electric current is passed through them. The drum on the press is "conditioned" with a special oil coating prior to the injection of ink on the drum. A writing head with an array of negative electrodes (200 per inch) passes pulses of electric current through the ink to release droplets of coagulated ink, and the drum serves as the positive anode.

The image is then "revealed" as the ink that is not adhering due to partial coagulation is scrapped off the drum with a rubber blade analogous to the doctor blades used in gravure printing. The image is then transferred to the paper under pressure and dries quickly.

The ink contains polymer strands, like plastic, that are suspended in a water solution. Different pulse durations are used for different areas on the page. For black ink alone, the cathode is on for about 4 seconds. For the small dots in highlight areas, the pulse is only about 100 nanoseconds. Varying the current produces tints or the grays. It is possible to control the thickness and size of the dots, which lead some people to claim that it is a continuous-tone process.

A buffer, called the "double buffer memory," allows the pages that will be printed next to be stored. There is a high-speed connection between the buffer and print heads that allow a transmission rate of 64 MB/sec.

With a resolution of 200 dpi, 8 bit/pixel, there is a theoretical possibility of printing 16 million colors. According to the manufacturer, to reach comparable image quality with offset technology would require 3200 dpi. However, we have heard similar claims in the past. When the distributors started to sell the Xeikon-based engines they claimed that the 600 dpi/color gave the appearance of 2400-dpi output. This, of course, is untrue.

Elcorsy estimates that the total printing cost, including equipment depreciation, paper, ink, and overhead, will be about 3.5 cents per page. We look at this product and the technology as neonatal and don't predict commercial product availability for several years.

# Chapter 34

# Comparisons

Using our definition of on-demand printing, several categories of equipment can be identified. Which part of the market you are from may determine what you call a demand press and which short-run printing product best fits your needs. For example, if you work in a large corporation with a mainframe you might consider the Delphax, Siemens, and Pennant demand printers. If you are a Fortune 500 company with an in-house reprographics department you could consider your DocuTech or Lionheart as an on-demand printer. If you are from the printing industry, you would consider your duplicator or Heidelberg presses (GTO-DI and QuickMaster) a demand press. If you are a corporate MIS center, a high-speed laser printer would be a typical system, and if you own a commercial prepress trade shop, you might consider your Indigo, Xeikon, and Chromapress an on-demand press.

Some organizations will install a variety of on-demand systems to meet a variety of needs. No one system appears to have the blend of features to meet all definable needs.

Regardless of what you or your peers already own or call an on-demand or digital press, the most difficult aspect if you are shopping for a unit is comparison. A full-fledged head-to-head comparison of all devices would be a book in itself.

## Printing and Duplicating

Printing and duplicating are terms that are used synonymously because they both deal with the replication of pages or documents. The distinction must be made, however, between conventional ink-on-paper printing that requires a one-time-use image carrier (plate) and toner-on-paper printing that applies a re-useable image carrier (drum or electronic plate).

Duplicating, as a term, was initially applied to lower-level printing presses (AM Multigraphics, A.B. Dick, etc.) that produced moderate quality black-and-white printing, and occasionally spot color. However, some people refer to higher-speed copiers as duplicators. These were commonly installed in quick printing and in-plant printing facilities. The formal differentiation between a duplicating press and conventional press is that the conventional press will have bearers, which are rings on the outside of the blanket cylinder and plate cylinder. The duplicator press may not.

The chart on the next page characterizes most of the replicating devices that are available. It contains a category called "Copier/Duplicator," which is the slang reference to copiers, not the true offset duplicator.

| | Speed | Quality | Page Size | Input | Carrier | Color |
|---|---|---|---|---|---|---|
| (Office) Printer | Low | Moderate | Standard | Data | None | Low–Mod |
| (Color Office) Printer | Low | Mod–High | To 11×17 | Digital | None | Low–High |
| Copier | Moderate | Moderate | To 11×17 | Analog | Drum | Low–Mod |
| Copier/Duplicator | High | Moderate | To 11×17 | Analog | Drum | Low |
| Printer/Duplicator | Mod–High | Moderate | To 11×17 | Data/Digital | Drum | Low–Mod |
| Digital Color Press | Moderate | Mod–High | To 11×17 | Data/Digital | Drum | Mod–High |
| Color Proofer | Low–Mod | High | To 11×17 | Digital | Drum | High |
| (Offset) Duplicator | High | Moderate | To 11×17 | Analog | Plate | Low–Mod |
| (Offset) Press | High | High | Signature | Analog | Plate | High |
| Direct to Plate on Press | High | High | Signature | Digital | Plate | Mod–High |

**KEY**

Speed:
- Low: 10 standard pages per minute or less, single sided.
- Moderate: 20–60 standard pages per minute, single sided.
- High: 90 pages per minute or more, single sided.

Color:
- Low: Black plus one color, or low-quality "process" color
- Moderate: Black plus three colors, or moderate-quality "process"
- High: High-quality "process" color—three primaries plus black

Input:
- Analog: A physical sheet of paper or some other substrate with images thereon, serving as a master image for reproduction. Requires platemaking for conventional printing.
- Digital: Electronic information consisting of rasterized text and images.
- Data: Text oriented.

## Copier Duplicators vs. Offset

Offset duplicators normally print on 8.5×11-in. (216×279-mm), 11×17-in. (279×432-mm), and related size sheets as do most copiers. There is strong competition between the two processes regarding speed, quality, and economics, and there are also offset presses in that size range. When comparing equipment, there are technical differences such as the number of form rollers, registration system, and other factors.

Except for offset printing personnel, most people only need to know that the press is capable of printing much higher quality on complex work. We believe that a press is capable of producing higher quality than any copier-duplicator, including digital.

For most, the underlying technology is not important. The following identify some of the more important factors to consider when comparing the two and some of our conclusions.

Quality is an obvious concern. Solids should be smooth, halftones should reproduce well, there should be fine line detail, the ability to hold faint images, and general overall quality. Ideally, this comparison should be made several weeks after the last maintenance. Both types of equipment may be excellent at the time of maintenance, but what are their relative qualities a short time later?
- Either process will do an excellent job on general linework (type, graphs, etc.) with the exception of letterhead, where the offset process is more professional.

- Digital copiers, particularly 600 dpi, will be superior to average offset duplicator work. However if the offset duplicator operator is highly skilled, then we prefer offset duplicator work. Analog copiers are third.
- Copiers can print solid coverage over a much larger area because they are not limited by the ink roller constraints of the offset process. Appearance over smaller solid areas is a matter of preference.
- Copiers are the clear preference on combination jobs in which a heavy solid is adjacent to a fine line. To the copier, "it doesn't matter." Offset duplicators have problems balancing the two.
- Operator skill required is an obvious concern. Operator skill requirements are greater for the offset duplicator, but the high-speed copiers with their many capabilities also require considerable skills.
- In looking at paper, the offset technology offers a clear advantage, since offset duplicators can run lighter and heavier stocks.
- The machine costs depend on the quantity and number of pages. Specifically the offset process has higher initial costs but lower per-copy costs. But as we demonstrated in earlier chapters, eventually at a certain quantity, offset printing becomes more economical.
- The operator costs generally favor the copier because the lesser skill requirements usually mean lower wages. Perhaps of greater importance is that it is common for one copier operator to run two copiers. We have seen one operator run three machines, but productivity will be lost unless two of the machines don't require much operator intervention and the copiers are laid out properly. With smaller presses such as offset duplicators, it is often one person per machine. It is possible to have one person run two, but that is unusual.
- In terms of speed, the copier is faster for short runs because no printing plate has to be made. Therefore, the copier can have the first copy reproduced in seconds.

Some of the automated in-line "paper master to offset duplicator systems" are not too far behind, but when making a printing plate that requires a negative, then even a person dedicated to the task will require close to 30 minutes to produce the first copy.

Also important is that the high-volume copiers have an interrupt feature which requires minimum time and bother to produce another job, even if a different paper is involved. Under the same situation, the offset duplicator will encounter considerable problems, including putting the original job's printing plate back on the offset duplicator.

## Comparison

If both are running 8.5×11-in. (216×279-mm) sheets with copy only on one side, then it is speed versus speed. The fastest offset duplicator is faster than the fastest copier.

The fastest high-speed copiers normally will outproduce the typical offset duplicators as long as both are running 8.5×11-in. (216×279-mm) sheets, black and white, one side. If any deviations occur and the offset duplicator has the capabilities, the offset duplicator will outproduce the copier.

The reason is that any deviation from an 8.5×11-in. (216×279-mm) sheet has a major impact on copier speed, but minimal or none on the offset duplicator. For example, a very high speed copier may have a rated speed of 8,100 copies of 8.5×11-in. one-sided sheets per hour, but if you change that to 11×17-in. (279×432-mm) sheets, copied two sides, the rated speed drops to 1,740 sheets per hour.

A duplicator with 11×17-in. (279×432-mm) duplexing capabilities will produce at the same speed and likely average 6,000+ 11×17-in. sheets per hour. Speed then must be based on quantity plus sheet size and number of copies.

## Copier vs. Offset Cutoff Points

The offset process requires the making of a printing plate, which can vary in costs, including labor:

- From less than 25¢ when making a paper master plate in a process similar to making a copy (some copiers can make paper masters, and some copiers can also be used to make a paper master plate for subsequent printing).
- Through several dollars for a direct-image plastic plate.
- To possibly more than ten dollars for a negative and a metal plate.

Copiers do not need an intermediate step to produce copies. At this point, copiers have a decisive cost advantage. Once the duplicator runs though, printing ink and chemistry will cost less than 10% of copier ink and chemistry (toner) costs, and has no meter click charges as most copiers plans have.

Simply put, offset duplicators cost considerably more to get the first copy, but each additional copy has a lower unit cost than the copier. The basic formula for one sheet (one or two pages) is:

$$\frac{(\text{Offset Pre-Run Costs} - \text{Copier Pre-Run Costs})}{(\text{Copier Running Cost} - \text{Offset Duplicator Running Costs})} = \text{Breakeven Point}$$

It follows then that the breakeven point would be directly related to the platemaking processes. For example:

- Offset duplicator pre-running costs are $0.24, including paper master cost, transfer of the master to the plate cylinder, and an average of three sheets of paper before the first good copy is achieved.
- Copier/duplicator scans in the copy when the copier is running, which is no pre-running cost.
- Offset duplicator has ink and chemistry costs of $0.0002.
- Copier has chemistry and meter click total charges of $0.0062.

The cutoff point calculation would be:

$$\frac{(\$0.24 - \$0.00)}{(\$0.0062 - \$0.0002)} = \frac{\$0.24}{\$0.0060} = 40 \text{ copies}$$

Forty copies is the breakeven point. More than forty copies can be economically produced on the offset duplicator. Less than 40 copies belongs on the copier. This, of course, is provided that either option can meet the quality and service requirements without any overtime.

Interestingly if only the metal plate option exists, the formula is not necessarily applicable. For example, suppose the metal plate option costs $10. Using the formula per se with the same running costs differences produces: $10/$0.006 = 1,667 copies.

Conceptually then, printing becomes more economical at 1,667 copies. In reality, if the film and metal plate costs $3.00 and no overtime is involved, then the thinking may (and likely should) be in terms of real dollars leaving the organization. This would equate to: $3.00/$0.006 = 500 copies.

If it is 1,667 copies, then $10 leaves the organization and goes to an outside copier vendor. If the offset duplication option is employed, then $3 (actually $3.33) goes to outside vendors with the remaining $7 staying within the organization.

Is 500 copies or 1,667 copies the correct break-even point? Most copier vendors will tell you that 1,667 copies is correct because they would treat internal and external money the same. We believe that they are quite different and as a result would lower the break-even point according to workload circumstances at the time.

If the operation typically produces 50 or more copies of a single sheet, it is worthwhile to investigate copiers that use intermediate masters. In essence, the machines have an initial master cost, but the per-copy costs should be lower than any copier option. The machines employ digital technology and are PostScript compatible. They fit in the office environment, where printing presses clearly do not, with speeds comparable to the fastest copiers and monthly ratings of 500,000+ copies. Duplexing is manual.

## Multiple-Page Copier Versus Offset Cutoff Points

When the economics of multiple-page work are considered, formulas still apply, but they must be coupled with common sense. In the previous section, the example paper master cutoff point equated to 40 copies for single-sheet work. Once multiple pages are involved, then the types of binding and the available equipment must be considered. For example:

1. If you deliver 8.5×11-in. (216×279-mm), flat, collated sets, such as loose-leaf or comb binding, and the printing option is coupled with sufficient sorter bins, then the primary difference is pulling out of many bins versus one copier location. The printing option has a minimal time and inconvenience increase over the copier. The cutoff point may increase to 50 copies. If stapling is involved, this must be done off-line with the printing operation, but on-line with the copier option. Dependent upon plant layout, the additional time may increase the cutoff point to 75 or 100 copies.

2. If the job was the same as the job above, but all collating was done off-line, then the decision is one of collating capabilities. On many collators, the job can gather more than one set per cycle, dependent upon the number of pages and pockets available. For example, if the collator has 50 pockets and the job has 14 sheets, it may be able to gather three sets (50/14) with each cycle.

3. If the job is saddle-stitched, some copiers can do the job on-line, while others require an off-line operation. The key considerations are number of pages and quantity. A press that perfects 11×17-in. (279×432-mm) sheets will require 20 to 30 minutes for each four-page form, dependent upon camera size, plate-making capabilities, and operator skills. Material costs must be added to this

time. Once running, the press is producing approximately four times as many sheets per hour as the high-speed copier and at a lower per-copy cost, even though copier companies generally treat 11×17-in. sheets as one meter click (same price as 8.5×11-in. images).

Another consideration is that the copier will saddle-stitch at full machine speed, so it makes no difference to the copier whether it is binding 8-page or 40-page booklets. The off-line saddle-stitcher may produce 50% more 8-page booklets per hour than 40-page booklets, but in terms of total pages, it is much faster with 40-page booklets, perhaps three or four times faster. Of course, if the booklet has a separate cover and the copier has no practical means of feeding the cover, then the copier option is likely out unless the quantity is quite small, and then manually inserting the covers may be a viable option.

Copiers generate heat. This will cause problems if inks such as rubber base are used because the ink will melt. Unlike the relative simplicity in determining the single-sheet cutoff point, multipage cutoff determinations require visualizing the procedures and costs of various alternatives.

## The Cost Factor

One area that we have not yet taken into consideration is the waste created as a part of the process or a by-product of inefficiency on the part of the system. For instance, the Xeikon roll of paper is "treated" or charged as the paper enters the imaging area. If the machine stops, the web of paper inside the machine looses its charge and is output blank. About ten pages are wasted. However, there is sufficient storage (a few gigabytes) for jobs to be queued for output, which would allow the machine to run continuously. However, there will be stoppages for any number of reasons.

Both Indigo and Xeikon are electrophotographic systems and cannot promise to deliver the same level of quality with every print. Some number will be unusable and thus wasted. For Indigo it might be as high as 20%; and for Xeikon, 10%, but these are very rough estimates and were not taken into consideration in our cost factors.

## Consumables

The table on the following page shows the consumable items for both systems. The data was acquired from cost figures provided to actual and potential purchasers. While Indigo has the attributes of a printing press with plate, blanket, and ink, it adds a special oil that is used with the ink. The Xeikon system is more like a copier or electronic printer. It has toner and the developer that works with it as well as an organic drum and accompanying brushes and scrapers for imaging pages.

We were surprised not to see corona wires on the list of consumables since these are common to the electrostatic process. As you can see, all of these items have different life expectancies — and even these are estimated. Thus, we must warn you that cost per-copy data derived from these numbers is not reliable.

## The Coverage Problem

Even in traditional printing, the amount of ink coverage based on the page image or images is considered in developing cost and price information. Digital printing must have the same consideration. We created two typical scenarios: 30% and 60% coverage of a standard 8.5×11-in. page. We then took an average of the two as a fair estimate and doubled it for the back side — we are now using duplexed pages as our standard.

The following are the major coverage considerations:
- Text coverage
- Image coverage
- Color coverage
- Combination coverage

Our 30% and 60% estimates are designed to fairly represent combination coverage over a wide variety and quantity of pages. As time provides more experience, we will modify our numbers.

## Consumables for Selected Digital Color Presses

The following data has been acquired from user feedback. Volume discounts and changing pricing make these numbers questionable, and they should be used with care. Ask your supplier or potential supplier for detailed pricing, and the coverage assumptions used. This pricing was in effect when the first edition of this book was published in September 1995.

**INDIGO**

| Unit | Impressions | Price | Cost/Page 4 over 0 30% coverage | Cost/Page 4 over 0 60% coverage | Cost/Page 4 over 4 Average |
|---|---|---|---|---|---|
| Blanket | 60,000 | $1,000 | $0.02 | $0.03 | |
| Plate | 30,000 | 700 | 0.02 | 0.054 | |
| Oil (/can/color) | 5,000 | 15 | 0.01 | 0.02 | |
| Ink (/can/color) | 3,000 | 115 | 0.15 | 0.31 | |
| | | | $0.21 | $0.41 | $0.62 |

**XEIKON (CHROMAPRESS, etc.)**

| Unit | Impressions | Cost/Page Price | Cost/Page 4 over 0 30% coverage | Cost/Page 4 over 0 60% coverage | 4 over 4 Average |
|---|---|---|---|---|---|
| Brush/scraper | 60,000 | $ 119 | $0.00 | $0.01 | |
| OPC drum | 150,000 | 560 | 0.00 | 0.01 | |
| Developer (/cartridge/color) | 50,000 | 464 | 0.04 | 0.07 | |
| Toner (/cartridge/color) | 15,000 | 280 | 0.07 | 0.15 | |
| | | | $0.12 | $0.24 | $0.36 |

Source: NEPP, *Digital Printing Report,* Vol. 1 No. 2, 1994. Information subject to change without notice.

The table below shows the Indigo consumables pricing in September 1996. Indigo provided us with current pricing levels for all E-Print consumables and clarified various points for us.

| Indigo | | | | 8.5×11 Yield @ 30% Coverage | 8.5×11 Yield @ 60% Coverage | 8.5×11 Yield Average Coverage | Cost 8.5×11, 4C, 1S | Cost 8.5×11, 4C, 2S |
|---|---|---|---|---|---|---|---|---|
| Consumable | Unit | 8-15-96 Price | 11×17 Yield | | | | | |
| Blanket | Blanket Cartridge | $550 | 60,000 | 120,000 | 60,000 | 90,000 | $0.006 | $0.012 |
| PIP (Plate) | Plate Cartridge | $380 | 40,000 | 80,000 | 40,000 | 60,000 | $0.006 | $0.009 |
| Imaging Oil | Bottle | $8 | 5,000 | 10,000 | 5,000 | 7,500 | $0.004 | $0.009 |
| ElectroInk | Canister | $32 | 3,900 | 7,800 | 3,900 | 5,850 | $0.022 | $0.044 |

The pricing for Electroink, for instance, has always provided discounts for quantity and payment. Thus most users were not paying the list price of $115 per can; most were receiving a 45% discount for certain volumes, which made the price $63.25. An incentive program for prompt payment offered two cans for the price of one, which brought the price down to $31.63. To simplify the entire process, Indigo set a street price of $39.58 to reflect what most users were already paying and then offered an incentive for prompter payment of 20%, which brings the street price to $31.67.

At the same time, Indigo increased the yield from plate and ink. Indigo now uses a different approach to forcing the ink out of the can. In addition its yield also increased for other more technical reasons. Indigo is using 3,900 11×17-in. sheets at 30% coverage for the yield, but this appears to be conservative.

## The Real Cost Problem

The cost of consumables is only the tip of the iceberg. In order to arrive at meaningful comparison data we have applied the budgeted hourly cost figures from the National Association of Printers and Lithographers (Teaneck, NJ) whose Blue Book series compares press and other printing equipment costs using a standard approach to developing reliable cost data.

That brings us to the following table, which was based on a printout from an Excel spreadsheet that compared people, overhead, equipment, depreciation, and consumables information.

We have tried to be scrupulously fair in developing this data, but because there are so many variables, we expect to reproduce this table many times. Note that we have assumed a full-time operator in some cases and a part-time operator for the more automated systems. It may be that this person is shared with other machines, but given the need to load and unload paper, send and queue jobs, and perform routine maintenance, we think it will be full time.

To arrive at a realistic comparison of "apples and apples," we have used an 11×17-in. sheet, four-color, duplexed, and folded. Because one system can fold at present

## Cost Comparisons for One-Page Flyer

### PER UNIT: 8.5×11 in., 4 colors, 1 side*

| | GTO 4C | GTO 2C | GTO-DI | QM-DI | O-A 5C | Indigo | Xeikon | CLC800 | CLC1000 | Docu-Color | Micro-Press |
|---|---|---|---|---|---|---|---|---|---|---|---|
| 50 | 3.33 | 4.46 | 1.04 | 0.86 | 0.98 | 0.08 | 0.09 | 0.18 | 0.19 | 0.19 | 0.15 |
| 100 | 1.67 | 2.23 | 0.52 | 0.43 | 0.49 | 0.08 | 0.09 | 0.18 | 0.19 | 0.19 | 0.15 |
| 500 | 0.34 | 0.45 | 0.11 | 0.09 | 0.10 | 0.08 | 0.09 | 0.18 | 0.19 | 0.19 | 0.15 |
| 1000 | 0.17 | 0.23 | 0.06 | 0.05 | 0.06 | 0.08 | 0.09 | 0.18 | 0.19 | 0.19 | 0.15 |
| 5000 | 0.04 | 0.05 | 0.02 | 0.01 | 0.02 | 0.08 | 0.09 | 0.18 | 0.19 | 0.19 | 0.15 |

### PER UNIT: 8.5×11 in., 4 colors, 2 sides*

| | GTO 4C | GTO 2C | GTO-DI | QM-DI | O-A 5C | Indigo | Xeikon | CLC800 | CLC1000 | Docu-Color | Micro-Press |
|---|---|---|---|---|---|---|---|---|---|---|---|
| 50 | 6.48 | 8.93 | 1.95 | 1.75 | 1.92 | 0.24 | 0.20 | 0.87 | 0.38 | 0.38 | 0.63 |
| 100 | 3.25 | 4.48 | 0.99 | 0.89 | 0.98 | 0.24 | 0.20 | 0.87 | 0.38 | 0.38 | 0.63 |
| 500 | 0.70 | 0.95 | 0.23 | 0.21 | 0.23 | 0.24 | 0.20 | 0.87 | 0.38 | 0.38 | 0.63 |
| 1000 | 0.35 | 0.48 | 0.13 | 0.12 | 0.13 | 0.24 | 0.20 | 0.87 | 0.38 | 0.38 | 0.63 |

*Paper and RIP time not included.

Source: NEPP, *Digital Printing Report*, Vol. 3 No. 5, 1996. Information subject to change without notice.

(Indigo) we have estimated a binding charge for folding as well as the re-running of sheets for systems that do not duplex (the GTO, GTO-DI, and the Canon) and trimming where needed by systems that print an oversized sheet (GTO and GTO-DI).

Paper costs were not included. This was one variable we were not prepared to deal with at this time. Our goal was to arrive at a "four-pager" cost per unit as well as a standard page — both duplexed four-color. To do so required some assumptions. Here are some of them:

- 80% productivity
- No waste
- One shift
- 10-year depreciation
- Maintenance contract for all
- "Traditional" overhead
- Ripping time off line
- No page creation costs
- No separations cost

As expected the conventional printing had to absorb film, plate, stripping (minor), and 54 minutes of makeready costs, but it had the highest speed. The GTO-DI had plates and 20 minutes of makeready (using the new high-quality Pearl laser technology), and almost the same speed — we slowed it down.

The Indigo and Xeikon systems have some interesting differences that can be attributed to the rather severe cost difference that resulted:

- The Indigo system is priced higher than the Xeikon, at least as present costs indicate.

- The Xeikon runs faster than the Indigo for duplexed, four-color printing.
- No attempt was made to deal with multipage booklets where the Indigo would have a clear advantage because of its integrated binding system.

The first thing that hits you is the production and economic efficiency of the conventional printing process: as run length increases, the cost per unit decreases. However, with purely digital printers the cost per unit is the same for any quantity. As we have said, these new printers will be used by commercial printing companies because of the synergy between conventional and new technology.

The Xeikon appears to be a clear winner over the Indigo based on cost, but Indigo quality is superior. But the jury is still out until we have actual user data. The Quickmaster-DI is very cost-effective but offers no variable data.

Keep in mind that the comparison is for duplexed sheets printed four-color. We can possibly draw some other conclusions:

- Conventional printing becomes competitive at about 5,000 copies. Although this comparison does not show it, multipage documents printed as signatures would probably bring the number down to under 2,000, especially with newer automated presses.
- The Quickmaster-DI is almost as cost-effective for short runs as the purely digital devices; yet, it can hold its own at longer runs. Remember that the DI is waterless offset and, without sufficient air conditioning, runs may have difficulty at 10,000 or more.
- The Canon does not print duplex, although the Xerox 5775, which would have similar numbers, will print some volume by re-running copies. Our numbers show a Canon copy generated from a RIP (as opposed to a copy generated from the scanner) at about $0.55 for one side, four-colors, 8.5×11 in. (216×279 mm). These are generally sold for about $2.00 each in tens and about $1.00 each in hundreds.

Let us keep in mind that these new systems are being designed to deliver finished publications (brochures, booklets, etc.) rather than flat sheets. And cost is only the beginning of price determination. The following table compares the speed of the Indigo E-Print and Xeikon-based devices.

|  | Indigo | | Agfa/IBM/Xeikon | |
| --- | --- | --- | --- | --- |
|  | Per Hour | Per Minute | Per Hour | Per Minute |
| 11×17 (A3) 1S, b/w | 4,000 | 66.7 | 1,050 | 17.5 |
| 11×17 (A3) 2S, b/w | 2,000 | 33.3 | 1,050 | 17.5 |
| 11×17 (A3) 4-over-0 | 1,000 | 16.7 | 1,050 | 17.5 |
| 11×17 (A3) 4-over-4 | 500 | 8.3 | 1,050 | 17.5 |

Double any of the above numbers to get 8.5×11-in. (216×279-mm) page speeds. Because the Xeikon engine has four duplex printing heads, there is no speed difference between single or duplex, one- or four-color printing, since the web moves at the same speed past all heads. The table on the next page compares major systems.

|  | **Indigo** | **Xeikon** | **GTO/DI-QM/DI** | **XDSS\*** |
|---|---|---|---|---|
| Imaging | Laser | Laser diode array | Laser | Laser |
| Max. sheet size | 13×18 | 12.6×17.9 | 14×20 | 12×18 (bypass) |
| Image area | 11×17 | 12.1×17.7 | <14×20 | 11×17+ |
| Paper | Sheet | Web/sheeter | Sheet | Sheet |
| Ink/toner | ElectroInk | Toner | Waterless ink | Toner |
| Resolution | 800 dpi | 600 dpi | 2,540 dpi | 400 dpi |
| Technology | Offset electrophoto. | Electrophoto. | Offset | Electrophoto. |
| Carrier | Reusable plate | Cylinder | Dry offset plate | Drum |
| 4C-over-4C/A3/hr. | 500 | 1,050 | 8,000–10,000 | 900 |
| Bleed | Yes | Yes | Yes | Yes (on 12×18) |
| Sides | Duplex | Duplex | 4-over-1 and single | Duplex |
| Variable data | Yes | Yes | None | Yes |
| Base/average price | $350,000– $450,000 | $350,000– $420,000 | $487,000– $600,000 | $160,000– $200,000 |

\*Xerox DocuColor/Scitex Spontane

**4-over-4, 8½×11 per hour**

| | |
|---|---|
| Agfa Chromapress | 2,100 |
| Canon CLC-1000 | 840 |
| Heidelberg GTO-DI | 2,660 |
| Heidelberg QuickMaster-DI | 5,976 |
| Indigo E-Print 1000 | 1,000 |
| IBM InfoColor 70 | 2,100 |
| Scitex Spontane | 900 |
| Xerox DocuColor | 900 |

Pricing data is estimated since various options are not yet formally priced, and early adopter and quantity pricing may not be typical. Also, consumables pricing is still evolving.

We recently visited three Indigo users on the West Coast. Two had two machines, and one had one machine. All had published price lists, so everything we are presenting is available to virtually anyone. There are no great secrets here. The oldest site we visited had had their machine for six months. So we were truly visiting the pioneers in digital color printing.

The table on the next page summarizes end-user prices in three run-length categories and averages them for simplex and duplex four-color printing on standard and tabloid sheets. The companies are listed as A, B, and C so you can see how their prices compared within the various categories.

| | 8.5×11, 4-over-4 | | | 11×17, 4-over-4 | | |
|---|---|---|---|---|---|---|
| | *50 - Total* | *50 - Each* | *Average* | *50 - Total* | *50 - Each* | *Average* |
| A | $200 | $4.00 | | $300 | $6.00 | |
| B | $178 | $3.56 | | $252 | $5.04 | |
| C | $225 | $4.50 | $4.02 | $325 | $6.50 | $5.85 |
| | *100 - Total* | *100 - Each* | *Average* | *100 - Total* | *100 - Each* | *Average* |
| A | $340 | $3.40 | | $442 | $4.42 | |
| B | $246 | $2.46 | | $377 | $3.77 | |
| C | $325 | $3.25 | $3.04 | $525 | $5.25 | $4.48 |
| | *500 - Total* | *500 - Each* | *Average* | *500 - Total* | *500 - Each* | *Average* |
| A | $640 | $1.28 | | $ 999 | $2.00 | |
| B | $840 | $1.68 | | $1095 | $2.19 | |
| C | $700 | $1.40 | $1.45 | $1050 | $2.10 | $2.10 |
| | 8.5×11, 4-over-0 | | | 11×17, 4-over-0 | | |
| | *50 - Total* | *50 - Each* | *Average* | *50 - Total* | *50 - Each* | *Average* |
| A | $150 | $3.00 | | $175 | $3.50 | |
| B | $100 | $2.00 | | $139 | $2.78 | |
| C | $175 | $3.50 | $2.83 | $225 | $4.50 | $3.59 |
| | *100 - Total* | *100 - Each* | *Average* | *100 - Total* | *100 - Each* | *Average* |
| A | $200 | $2.00 | | $260 | $2.60 | |
| B | $140 | $1.40 | | $211 | $2.11 | |
| C | $225 | $2.25 | $1.88 | $325 | $3.25 | $2.65 |
| | *500 - Total* | *500 - Each* | *Average* | *500 - Total* | *500 - Each* | *Average* |
| A | $467 | $0.93 | | $625 | $1.25 | |
| B | $495 | $0.99 | | $645 | $1.29 | |
| C | $480 | $0.96 | $0.96 | $630 | $1.26 | $1.27 |

Source: NEPP, *Digital Printing Report*, Vol. 1 No. 2, 1994. Some of the consumables are not yet meeting their run lengths claims. Information subject to change without notice.

Only one company had a minimum charge: $150. You can see from the chart that even the smallest jobs of 50 run length are at or above this amount. Two of the three had standard turnaround times of 48 hours. The third was 24 hours standard. One of the 48-hour services had a rush charge of +50% for 24 hours and +100% for 12 hours. No while-you-wait amounts were available. Almost all the users had selected Warren Lustro gloss coated stock in 80-lb. and 100-lb. text and cover weights. A version of this paper is made for the Indigo printers and has "Sapphire-like" characteristics.

## Comparison

The following chart depicts one Xeikon user's pricing for the categories shown above for Indigo users. We thought it would be interesting to compare. It is assumed that Agfa Chromapress users may be in the same pricing ballpark, but we will confirm that as we visit them over the next few months. We should note that the Xeikon users cannot handle the 100-lb. gloss coated stock.

| 8.5×11, 4-over-4 | | 11×17, 4-over-4 | |
|---|---|---|---|
| *50 - Total* | *50 - Each* | *50 - Total* | *50 - Each* |
| $96.00 | $1.92 | $166.50 | $3.33 |
| *100 - Total* | *100 - Each* | *100 - Total* | *100 - Each* |
| $120.00 | $1.20 | $220.00 | $2.20 |
| *500 - Total* | *500 - Each* | *500 - Total* | *500 - Each* |
| $485.00 | $0.97 | $895.00 | $1.79 |
| **8.5×11, 4-over-0** | | **11×17, 4-over-0** | |
| *50 - Total* | *50 - Each* | *50 - Total* | *50 - Each* |
| $69.00 | $1.38 | $112.50 | $2.25 |
| *100 - Total* | *100 - Each* | *100 - Total* | *100 - Each* |
| $100.00 | $1.00 | $180.00 | $1.80 |
| *500 - Total* | *500 - Each* | *500 - Total* | *500 - Each* |
| $400.00 | $0.80 | $725.00 | $1.45 |

You can clearly see that the Xeikon pricing is less than the Indigo pricing. However, we discovered that price really is not the issue in digital color printing. Turnaround time is. That's what users are buying. Much of what we saw in terms of work included posters, presentation sheets, special reports, sales materials, and various kinds of signs. Where the Xeikon may have a clear advantage is in short-run books. The West Coast users naturally had unique jobs that East Coast users might not see. One was an extensive folder of materials to hype a film or TV show. The required amount was less than 200, but the kit had folders, cards, sheets, and other formats within it.

Most of the users had at least a cutter. All had a relationship with a local bindery. What all the users were concerned about was competition from the next wave of users. Would they arbitrarily lower prices to grab market share and thus cannibalize the market? And what effect might the new Canon CLC-X and the Scitex Spontane have? These early users had invested almost $1 million to be first in digital color printing. They were seeing business from their existing customer base — all were and still are prepress services — and new customers. All had separated the new company under a different name from their existing business. All were very positive about their supplier, their future, and their success.

# Chapter 35

# Summary

On-demand digital printing is a moving target. The definition today is different than it was two years ago, and it may be different in two more years. As we have discussed throughout this book, the definition today is a combination of three things:

- Old and new technology
- Markets and applications
- Sales and marketing

## Old and New Technology

Today on-demand digital printing is both old and new technology. It is the old black-and-white copy technology used in the DocuTechs, it is the new color technology used in the color copiers and Xeikon machines, and it is the new hybrid technologies used in the Indigos as well as the digital Heidelberg.

Tomorrow the technology could include:

- XMX and Field Effect Imaging
- MAN-Roland Digital Change-Over Web
- Scitex high-speed inkjet
- Workgroup technology from Delphax, Siemens-Nixdorf (now Océ), Datametrics
- Elcography

It's not just output technology either. The printing portion is surrounded by the file transfer and the technology to finish the product. The inability or time delay to get paper or a delay in transferring the file can create a problem. And the inability or time delay to perform the binding and finishing can be a problem.

## Markets and Applications

While few people want to admit it, it appears that some of the market research or at least the market acceptance is less or slower than originally predicted. While many felt the market would have matured by 1997, it appears that the market will not become a mature market until approximately the year 2000.

Successful on-demand and digital color printers understand that it is how you position your products and services and how you satisfy your customer. It is using the right technology for the right applications. It is just-in-time delivery where and when the product is needed. Remember that the four advantages of on-demand digital printing are:

- Reduced cycle time
- Shorter run length
- Lower cost per copy of short runs
- Variable printing

Last but certainly not least, remember the four points about selling:
- **Product positioning:** don't position it as the same old thing
- **Learn and promote the benefits:** "This is not your father's car"
- **Reconsider your contact:** that old person may be the wrong person
- **Think creatively:** figure out clever ways to sell this service
- **Overcome buyer resistance:** push convenience; build confidence

## Demand Printing Is Not a New Idea

Demand printing is a moving target. It is technology and application. It is old technology and new technology. It's how you satisfy your customer. It is just-in-time delivery where the product is needed and when the product is needed. It should be cost-effective. On-demand is a philosophy, and you can make it whatever you want it to be.

If we were to summarize the most salient points about on-demand printing, they would be:
- There is a definable market for on-demand printing.
- There is a growing desire for more color.
- There is an increasing amount of technology being applied.

From our expanded definition of on-demand printing to our review of published market research, we feel more confident that our industry is on the right track to meeting the needs of customers in new and innovative ways.

## Available Applications

Virtually every category of reproduced product is applicable for on-demand approaches. Some categories have immediate application while others will develop over time. Through case histories we have tried to show how users of on-demand approaches are building markets.
- There is a market for black-and-white on-demand printing.
- There is a market for color on-demand printing.
- There is a market for both.

However, everyone now admits that the early adopters of digital color presses have not made any real money. Considering the size of the capital investment, the high cost per unit, and the need to develop a new market, it is no wonder. Although the so-called on-demand, short-run color market is admittedly large, the early adopters do not seem to have tapped into it as quickly as predicted.

If digital color printing services are to get any work, it will come from four primary areas:
- **Offset transfer:** work that was previously printed by offset but all or part of it converts to digital printing. This would be the major area of potential work. Since offset accounts for so much of our printing, it is only natural to attack it.

In fact, it was the cost structure of offset in the under 10,000 run that helped to define the market. Then offset became efficient at 5,000, and now it is efficient down to runs of about 3,000.

- **New applications:** work that was not printed before — or did not even exist before. Like TV tape recording and color copying, no one really knew there was a definable market there.
- **Color copier transfer:** work that was done on color copiers, but all or part of it converts to higher-capability digital color systems. Work here would include all sorts of longer and longer short runs. Digital color will begin where color copiers have heretofore left off: larger sheet sizes, bleed, duplexing, higher resolution, etc.
- **Black-and-white work done in color:** monochrome work that converts to color printing.

The impediments to growth have been:
- **Reliability:** Early systems had limited productivity. Many systems were introduced too early in their development cycle.
- **Cost per unit:** Capital and consumables costs were high to start, and that made the cost per copy high.
- **Other limitations:** sheet size and quality come to mind.
- **Speed:** More work produced per unit of time. That helps to reduce costs.

## Available Technology

Our review of available technology shows that many different technical approaches are being developed for this area, from changes to the traditional reproduction processes to new digital technologies. Each appears to have a niche where it has the advantage of cost-effectiveness. There are many technologies, and it will be a challenge to pick the one or ones that best meet the needs of the product mix.

- Computer-to-plate and on-press automation make traditional printing more effective.
- Electrophotography and inkjet hold great promise.
- Other technologies will find acceptance.

But technology comes at a price, especially when it is new. At a Seybold event in San Francisco, an early adopter of digital color printing came to the podium at one of the many sessions and presented a sincere picture of one year in the life of a "pioneer." What the person said will be lost in the plethora of presentations, since there were no visuals. The name of the user and the system are not provided, so that the information cannot be used by competitors, since much of what was said could be said about many of the vendors and systems. The points that were made do not require names associated with them:

1. No one was really ready for new technology. The system produced three pages for every one that was acceptable. The vendor did not understand the market. The user did not understand the market. The buyers of digital color printed pages did not understand the market. They were all learning together, but the user was paying for almost all of it. It took nine months for the learning curve

to level out, and this was with experienced color prepress professionals who, in our opinion, were very smart. Not all users are smart.

2. You have to make a profit. High-tech equipment must be written off over two years, and its return on investment must be fast and continuing. Why? So that you can afford the next level of technology. It took this user nine months to break into profitability. Then after three months of profit, the vendor introduced a new capability which is essential to the user. The price of that option wipes out the three-month profit. The user understood clearly that electronic technology is obsoleted at a rapid rate and you must make your money quickly to stay ahead of competitors who did not buy in at the early adopter stage but buy in later and reap the benefit of better systems, allowing them to compete strongly. Vendors should have an "Early Adopter Protection Program."

3. There are three levels of cost and only one level of profit. In addition to the capital cost, there is the cost per unit, which vendors and their consultants say does not matter, and maintenance cost. Our stalwart user likened it to sending a royalty back to the vendor with every page produced, good or bad. Vendors are addressing this with better yields, better pricing, and better maintenance plans. But for the early adopters, they had to struggle with high cost.

4. Quality is not the primary issue. What the systems produce is more than acceptable. The real issue is pricing. Buyers will pay a certain premium for turnaround time, just as they did with the first color copiers. They will pay a slight premium for on-demand printing. But every early adopter is paranoid about the second wave of systems: Xerox DocuColor, Scitex Spontane, and Canon CLC-1000. They all have a lower cost structure. Thus, their users may be apt to compete on price. Once that happens, every business model goes out the window.

5. Digital color printing is the future. There are still many issues to contend with, but the market is there and is slowly growing. This early adopter would probably do it again, but there has been much soul-searching over the last year.

## Digital Color Printing: Too Early or On Time?

At least two major players in digital color printing have stated that there was slower-than-expected development of the short-run color market. Let's recall what has happened so far:

- September 1991: Heidelberg announces GTO-DI at Print 91.
- January 1992: Heidelberg installs GTO-DI at Sir Speedy in LA.
- June 1993: Indigo announces E-Print 1000.
- July 1993: Xeikon announces DCP-1.
- September 1993: Both Indigo and Agfa show systems at IPEX 93 in Birmingham, England.
- Mid 1994: Less than 20 Indigo and AM (Xeikon) units in U.S.
- Early 1995: 400 Indigo units and 400 Xeikon units worldwide.
- May 1995: Drupa showcases digital printing as working technology.
- August 1995: Indigo upgrades all systems to improve reliability.
- January 1996: About 1,000 digital color printers and 110 hybrid (platemaker/presses) installed worldwide — all in less than 400 U.S. sites.

In other words this market is still very young — really less than two years old. The gap between the announcements and the installations has certainly been a factor in the development of this market. There are two requirements for a market to develop:

**1.** A large base of buyers.

**2.** A reasonably large base of sellers.

For buyers of digital color there is the need for digital document preparation, which means they must be users of desktop or other digital technology. Fortunately, 70% to 80% of all pages now printed are prepared digitally, so the base is large and growing. Unfortunately there were not that many sellers for some time, and the early adopters lived with reliability and quality problems that have since been pretty well corrected.

You must draw parallels from the development of color copying. From virtually zero in 1987 until today, color copying grew because there were many users and they were all promoting their services. The EFI Fiery RIP made color copiers *de facto* digital printers and over a third of all color copies go through the RIP, so there definitely is a digital printing market.

The Indigo, Agfa, AM, IBM, Xeikon, Scitex, and Canon printing systems are only now creating a large enough base for business. Now the real promotion and marketing must begin — to end users. With low-end desktop printers, mid-range color printers, and digital color presses, there are now sufficient alternatives for color printout. But suppliers and users alike now bear the burden of marketing their services. When nobody sells anything, a terrible thing happens: nothing.

## Conclusion

One way or another, we are all reproducing information on paper, on-demand for customers. We will apply many different devices and different technologies to accomplish this.

# Index

# About the Authors

**Howard M. Fenton** is a senior technical consultant/digital technology with the Graphic Arts Technical Foundation, Pittsburgh, Pennsylvania. The former editor-in-chief of *Pre* magazine, he is well known as a trainer, consultant, and speaker in print production. He has written over 200 articles, including a section in the *McGraw-Hill Science and Technology Yearbook* and scripts for the "Agfa Across America" educational seminars and PIA's PrintScan audiotape series. Fenton has also owned and operated a graphics company, managed a retail desktop publishing store, managed several prepress facilities, and taught advanced computer graphics at Montclair State College in New Jersey.

**Frank J. Romano** is the Melbert B. Cary Jr. Distinguished Professor of Graphic Arts at Rochester Institute of Technology, Rochester, New York. Romano teaches electronic prepress and electronic publishing at RIT's School of Printing Management and Sciences. The author of eleven graphic arts books, including GATF's *Computer-to-Plate: Automating the Printing Industry* book (with Dr. Richard M. Adams II) and the first book on QuarkXPress, Romano is also the founder of the QuarkXPress Users International Association. In addition, he serves as founding editor for the PennWell Graphics Group publications, including *Electronic Publishing,* which he founded in 1977 as *TypeWorld.*

*Contributor:*

**Frank V. Kanonik** is director of on-demand printing services for the Graphic Arts Technical Foundation. With 20 years of graphic arts experience, Kanonik has detailed knowledge of on-demand printing and electronic prepress in the Macintosh and Windows environments. He teaches numerous GATF workshops, in-plant training programs, and seminars pertaining to prepress technologies, both conventional and electronic, with a special emphasis on digital on-demand printing. Kanonik joined GATF as a black-and-white process photographer and later worked as a color camera operator and photographer for several companies in Florida.

# About GATF

The Graphic Arts Technical Foundation is a nonprofit, scientific, technical, and educational organization dedicated to the advancement of the graphic communications industries worldwide. Its mission is to serve the field as the leading resource for technical information and services through research and education.

For 73 years the Foundation has developed leading edge technologies and practices for printing. GATF's staff of researchers, educators, and technical specialists partner with nearly 2,000 corporate members in over 65 countries to help them maintain their competitive edge by increasing productivity, print quality, process control, and environmental compliance, and by implementing new techniques and technologies. Through conferences, satellite symposia, workshops, consulting, technical support, laboratory services, and publications, GATF strives to advance a global graphic communications community.

The Foundation publishes books on nearly every aspect of the field; learning modules (step-by-step instruction booklets); audiovisuals (CD-ROMs, videocassettes, slides, and audiocassettes); and research and technology reports. It also publishes *GATFWorld,* a bimonthly magazine of technical articles, industry news, and reviews of specific products.

For more detailed information on GATF products and services, please visit our website *http://www.gatf.lm.com* or write to us at 200 Deer Run Road, Sewickley, PA 15143-2328 (Phone: 412/741-6860).

# Other Books of Interest

The Graphic Arts Technical Foundation (GATF) is a major publisher of books on printing and related topics. Following are some of GATF's more popular titles:

- *Computer to Plate: Automating the Printing Industry* by Richard M. Adams II and Frank J. Romano

- *Understanding Digital Color* by Phil Green

- *Guide to Desktop Publishing* by James Cavuoto and Steven Beale

- *Understanding Electronic Communications: Printing in the Electronic Age* by A'isha Ajayi and Pamela Groff

- *Glossary of Graphic Communications* compiled by Pamela Groff

- *Professional Print Buying* edited by Phil Green

- *Handbook of Printing Processes* by Deborah Stevenson

- *The Magazine: Everything You Need to Know to Make It in the Magazine Business* by Leonard Mogel

- *Screen Printing Primer* by Babette Magee

- *Flexography Primer* by Donna Mulvihill

- *Lithography Primer* by David Saltman and Nina Forsythe